P9-EGM-025

Introductory Physical Science

 Seventh Edition

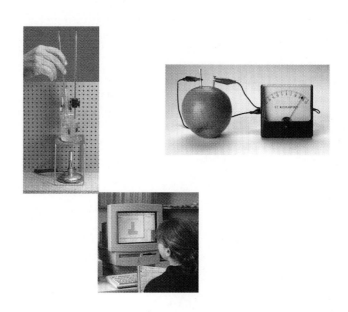

Uri Haber-Schaim

Reed Cutting

H. Graden Kirksey

Harold A. Pratt

Science Curriculum Inc., Belmont, Massachusetts 02478

Introductory Physical Science
Seventh Edition

Uri Haber-Schaim · Reed Cutting · H. Graden Kirksey ·
Harold A. Pratt

©1999, 1994, 1987, 1982, 1977, by Uri Haber-Schaim; ©1972
by Newton College of the Sacred Heart. Copyright assigned to
Uri Haber-Schaim, 1974; ©1967 by Education Development
Center, Inc. Published by Science Curriculum Inc., Belmont,
MA 02178. All rights reserved. No part of this book may be
reproduced by any form or by any means without permission in
writing from the publisher. This edition of *Introductory Physical
Science* is a revision under free licensing of the work under the
same title copyrighted originally by Education Development
Center, Inc. The publication does not imply approval or dis-
approval by the original copyright holder.

Credits

Editor: Sylvia Gelb

Book Design: SYP Design & Production
    Cambridge, MA 02139

Photography:

    Benoit Photography: pages 2, 4, 13, 15, 27, 39, 41, 46, 56,
    79, 89, 91, 101, 113, 117, 138, 141, 142, 151, 171, 180, 193,
    194, 195, 197, 198, 200, 201, 205, 207, 209, 216, 221, 227,
    229, 233, 235

    Douglas Christian: pages 3, 19, 20, 55, 80, 81, 119, 122

    Stephen Frisch/Stock Boston: page 213

    Fundamental Photographs: page 111

    Susan van Etten: page 65

    José Luis Banus-March/FPG International LLC: page 1

    Stevie Grand/Science Photo Library: page 135

Photo researcher: Susan van Etten

Manufacturing consultant: Mary Day Fewlass

Printed in the United States of America
by World Color

ISBN 1–882057–18–X

10 9 8 7 6 5 4 3 2

Preface to the Seventh Edition

The basic purpose of the course is to give all students a beginning knowledge of physical science and to offer insight into the means by which scientific knowledge is acquired. In pursuing this purpose, the *IPS* course helps students to understand some of the basic principles of physical science and to acquire useful laboratory skills; it also develops reasoning and realistic problem-solving skills. Furthermore, the course encourages communication by requiring individual students to take part in a cooperative learning process. *IPS* students learn from nature, the text, their teachers, and each other.

IPS implements the spirit of the National Science Education Standards by practicing inquiry throughout and avoiding a "mile wide and inch deep" syllabus.

The theme of the course is the development of evidence for an atomic model of matter. Rather than broadly surveying the entire field of physical science, we have taken a well-defined path toward this major objective. The method employed to achieve the stated goals is one of experimentation and guided reasoning based on the results of student experiments. Thus the body of the text includes laboratory experiments which students must carry out to understand the course properly. Many of the conclusions and generalizations arrived at and recorded in student laboratory notebooks complement the text in an essential way.

These attributes are as valuable for today's students, who will lead their adult lives in the twenty-first century, as they have been for the fifteen million students who used earlier editions of *IPS*.

Compared with the changes made in the Sixth Edition, the changes in the Seventh Edition are minor, but still significant. Even a quick perusal of the textbook will reveal a great improvement in the clarity of the photographs, especially those of equipment being used in experiments. A number of older black-and-white photographs have been replaced, as have several drawings and computer-generated histograms. The software to generate these histograms is now available and is recommended for use by the students.

Changes in content were made in Chapter 3, where a section on Boiling Point and Air Pressure has been added, and the section on The Hydrometer has been deleted. (Most car batteries are now sealed, making the illustration obsolete.)

In Chapter 6 the experiment on The Decomposition of Sodium Chlorate has been deleted. The Heating of Baking Soda in Section 1.1 gives students an experience similar to that gained from the Decomposition of Sodium Chlorate. An expanded introduction to Chapter 6 has further smoothed the flow of ideas.

In Chapter 9 the experiment on The Size and Mass of an Oleic Acid Molecule has been rewritten. The results are more convincing, as has been shown by field-testing. These changes in the text resulted in the addition and deletion of a few questions at the end of sections and chapters.

November 1998

Uri Haber-Schaim

Reed Cutting

H. Graden Kirksey

Harold A. Pratt

Contents

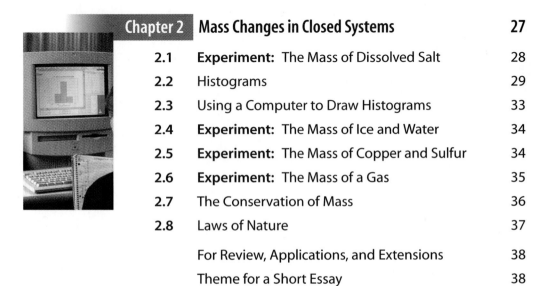

Chapter 5 — The Separation of Mixtures — 89

Chapter 6 — Compounds and Elements — 111

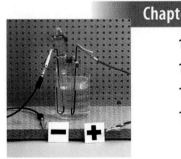

About the Course

The main topic of this course is the study of matter. This can be quite a challenge. Just look around you and note the large variety of substances and forms. There may be concrete walls, wooden desks, glass windows, and metal pipes. You can make the list as long as you wish.

We can develop ideas that explain the endless variety of material things in terms of fewer, simpler things. Many different things could conceivably be made from the same units by putting them together in a variety of ways. For example, bricks can be used in many ways—to make a wall or a house or a doorstop or a paved street. Over 2,000 years ago, the Greek philosopher Democritus conceived of small units that he called "atoms." But Democritus did not know whether atoms really existed or how many different kinds there were. His ideas must have been important, because we still use the word "atom." However, the word in itself does not really explain anything. It did not help people to predict any properties of matter or to understand what kinds of changes could take place.

If someone explained that a TV works because invisible gremlins paint the picture on the front of the tube, would you be satisfied? Using the word "gremlin" does not help you understand how a TV works. Nor does the statement "matter is made of atoms" help you much in understanding matter or atoms.

Modern chemistry and physics can give a much more meaningful account of the properties of matter. If this account is to have any meaning for you, we shall have to begin at the beginning. We cannot just throw new words at you. Each step must be filled in with experiments that you will perform. Then all the words and ideas will correspond to something real for you, and you will reach conclusions on your own.

We shall first concentrate on a few properties of matter that you can observe and measure. Then you will learn how to use them to distinguish between different materials. Knowing how to distinguish between materials allows you to separate and sort them by their properties. To organize this knowledge, we shall develop a model that you can use to predict the results of new experiments. This will enable you to put some real meaning into the word "atom."

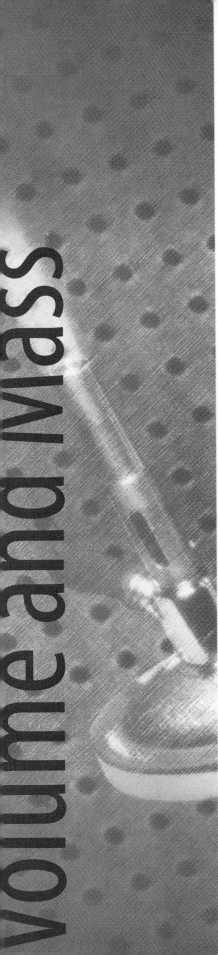

Chapter 1
Volume and Mass

There are many ways to begin the study of physical science. We shall begin with a simple experiment that produces some surprising results. The experiment will raise some questions. At the same time, you will learn very useful laboratory skills.

EXPERIMENT
1.1 Heating Baking Soda

What do you think will happen if you heat some baking soda in a test tube? Will the baking soda change color? Will anything leave the test tube? Before you read on, try to predict what will happen and be prepared to explain on what you base your prediction.

Put some baking soda into a dry Pyrex test tube to a height of about 0.5 cm. In case a gas is produced, it will be useful to be able to collect it. You can do that with the apparatus shown in Figure 1.1. The bent glass tube may already have been inserted into the rubber stopper. If you have to insert the tube yourself, look closely at Figure 1.2 and its caption for useful advice on how to do it safely.

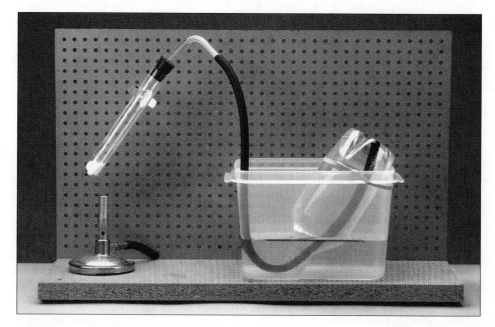

Figure 1.1
The apparatus used in heating baking soda. To be sure that no water will spill during the experiment, fill the container only up to the marker. Fill the bottle to the top and hold your hand over the mouth of the bottle while inverting it and placing it in the container. The rubber band helps to hold the bottle in place. Finally, insert the rubber tube all the way up to the bottom of the inverted bottle.

After your teacher has checked your setup, you may begin heating the baking soda over either a microburner or an alcohol burner. Look at Figure 1.3 to see how the flame on either burner should appear.

CAUTION: Always wear safety glasses when you use a burner or work with gases.

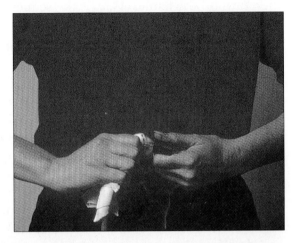

Figure 1.2
Lubricate the glass tube with glycerine or water before pushing it into the stopper. Hold the tube with a towel as close to the stopper as possible while pushing it in. This will prevent the tube from hurting you in case it breaks.

Figure 1.3
The proper size and color of the flame for a microburner (left) and for an alcohol burner.

Watch the test tube and collecting bottle as the baking soda heats up.

- What do you observe at the bottom of the test tube? Near the top?
- Does a gas collect in the inverted bottle?

When it appears that no further changes are taking place, pull the rubber tube completely out of the water while the flame is still on. Only then turn off the flame. (Why is the sequence important?)

CAUTION: Do not blow out the flame of a microburner; turn the valve at the gas jet.

Even without measuring, you can see that there is more gas in the bottle than there was air in the test tube. To make sure of this, we heated an empty test tube connected to the same collecting bottle shown in Figure 1.1. All the air that was driven out of the heated test tube shows up as the bubble shown in Figure 1.4.

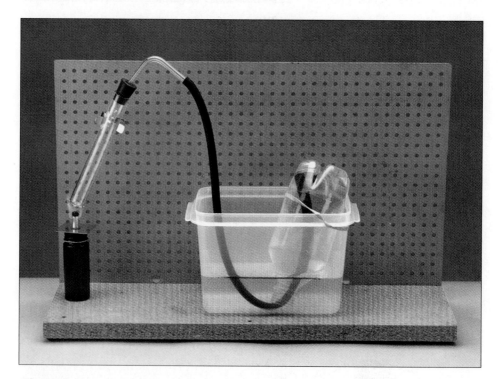

Figure 1.4
The result of heating an empty test tube for a few minutes. Clearly, the gas in your experiment was not just air from the test tube.

- From where, do you think, did the gas come?
- From where did the droplets on the test tube come?

As far as you can tell by looking, the baking soda in the test tube remained unchanged. But did it really? You are going to answer this question with a simple test.

First let the test tube in which you heated the baking soda cool down. Then put about an equal amount of baking soda in a second test tube. Pour some tea into each of the two test tubes (about a quarter tube will do). Gently shake both test tubes or use a stirring rod to dissolve the powders. You have treated the content of the two test tubes equally.

- Is the color of the liquid in the two test tubes the same?
- If the color is different, can the white powder in the test tube that you heated still be baking soda? Why or why not?
- How can you compare the amounts of solid, liquid, and gas that you observed?

One of the best ways to find out what a thing is made of—and how it works—is to take it apart. This is what you did when you heated baking soda. But when you take something apart, the manner in which you do so affects the results. Suppose you had thrown some baking soda into a flame. It would surely have heated up, but you would have learned very little. By heating the baking soda under controlled conditions, you were able to make some interesting observations.

Note that in the experiment you did, only the droplets on the upper end of the test tube were directly visible. Without displacing the water in the collecting bottle, you would not have detected the gas. Without using tea as an indicator, you would not have known that the white powder in the test tube was no longer baking soda.

In this experiment the collecting bottle and the tea served as tools to extend your sense of sight. The use of tools to extend the human senses makes advances in science possible. You will learn to use a variety of such tools in this course.

1. Why do you think baking soda is used in baking?

2. List some tools that you have used
   a. to extend your vision to see distant objects.
   b. to extend your vision to see very small objects.
   c. to tell how hot something is.

1.2 Volume

Suppose that you have some pennies stacked one on top of another in several piles, and that you want to know how many pennies are in each pile. The obvious thing to do is to count them. If you had to count the pennies in many piles, you could speed up the counting in the following way: Make a scale like that shown in Figure 1.5, marking it off in spaces equal to the thickness of one penny. You can then place this scale alongside each pile and read off the number of pennies.

If you want to measure the amount of copper in each pile of pennies, you first have to decide in which unit to measure the amount of copper. If you choose as the unit the amount of copper in one penny, then the amount in the whole pile is expressed by the same number as the number of pennies.

Suppose, now, that you want to find out how much copper there is in a solid rectangular bar of copper. You might think of making a box of the same size and shape as the copper bar and then counting the number of pennies needed to fill the box. This idea will not work, because if you place pennies next to one another in a rectangular box, there will always be some empty space between them.

A better way to measure the amount of copper in the bar is to choose a new unit, such as the volume of a small cube. Suppose you had a box the same size and shape as the copper bar, and you could fill it with cubes of copper of a size that would fit without air spaces between them. You could simply count the number of cubes to find the amount. Of course, you do not have to count each cube. If a cubes fit along the length of the box, b along the width, and c along the height, then the total number of cubes in the box (and bar) is $a \times b \times c$. (See Figure 1.6.) This is the amount of copper in the solid bar, expressed in units of cubes. As you probably know, this is also the volume of the bar, expressed in terms of the *volume* of the unit cube.

What we choose to be the length of each side of this unit cube is a matter of convenience. We shall choose a unit of length based on the meter (m), the international standard of the metric system.

Figure 1.5

A scale for counting the number of pennies in a vertical pile. The distance between marks is the thickness of one penny.

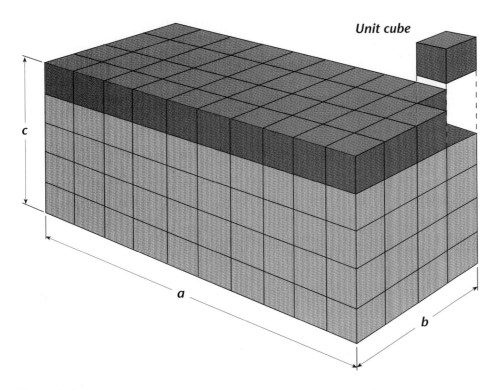

Unit cube

Figure 1.6

A bar of copper 10 cubes long, 4 cubes wide, and 5 cubes high. One layer of the bar contains 10 rows of 4 cubes each, or 10×4 cubes. There are 5 layers in the bar. Each layer contains 10×4 cubes. Therefore, the number of unit cubes in the bar is $10 \times 4 \times 5 = 200$. If the unit cube is 1 cm on an edge, the volume of the bar is 200 cm^3 (cubic centimeters). For any rectangular solid, therefore, the volume is the product of the three dimensions, $a \times b \times c$.

In this case, as in much of our work in this course, we shall use the *centimeter* (cm). A centimeter is 0.01 m. Our unit cube would then be the *cubic centimeter* (cm^3), a small cube 1 cm on an edge.

To sum up, then, we can compare different amounts of the same substance by comparing their volumes—that is, the amounts of space they occupy. For a rectangular solid, we find that volume by measuring its three edges and calculating the product of these numbers. The volumes of solids of other regular shapes can also be calculated from measurements of their dimensions, but this requires additional knowledge of geometry.

The use of volume to compare amounts of substances is particularly convenient in the case of liquids, because liquids take the shape of their containers. Suppose you wish to compare the amounts of water in two bottles of very different shapes. You simply pour the contents of each separately into a graduated cylinder that has already been marked off with the desired units, and read off the volumes (Figure 1.7). This way of measuring volume is very much like counting pennies all stacked up in a pile.

You can use the property of a liquid to take the shape of its container when you want to find the volume of a solid of irregular shape, such as a small stone. After pouring some water into a graduated cylinder and reading the volume of the water, you can submerge the stone in the water and read the combined volume of the water and the stone. The difference between the two readings is the volume of the stone.

Figure 1.7
A graduated cylinder marked off in units of volume. The cubic-centimeter marks could be made by filling the cylinder with liquid from a small cubic container, 1 cm on an edge, and making a mark at the liquid level each time a container-ful of the liquid is poured in. Many graduated cylinders are marked off in milliliters (mL). A milliliter is the same as a cubic centimeter.

†3. A student has a large number of cubes that measure 1 cm along an edge. (If you find it helpful, use a drawing or a set of cubes to answer the following questions.)
   a. How many cubes will be needed to build a cube that measures 2 cm along an edge?
   b. How many cubes will be needed to build a cube that measures 3 cm along an edge?
   c. Express, in cubic centimeters, the volumes of the cubes built in (a) and (b).

4. One rectangular box is 30 cm long, 15 cm wide, and 10 cm deep. A second rectangular box is 25 cm long, 16 cm wide, and 15 cm deep. Which box has the larger volume?

5. Figure A shows a cone-shaped graduate used for measuring the volume of liquids. Why are the divisions not equally spaced?

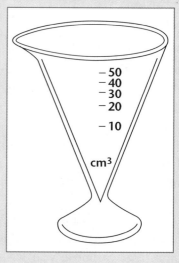

Figure A
For problem 5

†Answers to questions marked with a dagger are found on pages 250–51.

1.3 Reading Scales

To measure length with a ruler, volume with a graduated cylinder, and temperature with a thermometer, you must be able to read a scale. Therefore, learning how to get all the information a scale can provide is a useful skill.

We shall begin with reading a metric ruler (Figure 1.8). The smallest divisions on such a ruler are 0.1 cm (1 mm) apart. This is a small distance indeed. Nevertheless, when the object you wish to measure has sharp edges, you can see whether the edge falls on one of the lines.

In Figure 1.9, the edge falls between two lines. It is clear that the length is between 4.8 cm and 4.9 cm. To gain more information, you can estimate the position of the edge. If you cannot tell whether the edge is closer to one line or the other, it is best to report the reading as 4.85 cm, or 48.5 mm.

If the edge is closer to the line on the left, you should report the reading as 4.82 cm or 4.83 cm. Either way, you will not be off by more than ±0.02 cm. (The notation ± means "plus or minus.") Similarly, if you decide that the edge is closer to the line on the right, report the reading as 4.87 cm or 4.88 cm. Again, you will not be off by more than ±0.02 cm. Had you read the scale as 4.8 cm or 4.9 cm, you might have been off by as much as ±0.05 cm.

Figure 1.8
A metric ruler. The numbered divisions are centimeters.
The small divisions are 0.1 cm, or *millimeters.*

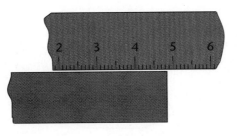

Figure 1.9
Reading the position of the edge of an object. Here the
edge falls between two of the millimeter marks.

Suppose that, as far as you can tell, the edge falls on a line (as shown in Figure 1.10). Then you should report the reading as 3.20 cm. This will indicate that the reading is closer to 3.20 cm than to either 3.22 cm or 3.18 cm. Here the "0" gives us information that would have been lost had you written only "3.2 cm."

Figure 1.10
Reading the position of the edge of an object. Here the edge falls on one of the millimeter marks.

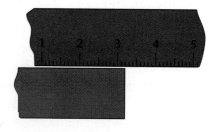

6. The scale in Figure B is in centimeters. Estimate the positions of arrows *a* and *b* to the nearest 0.1 cm. Can you estimate their positions to 0.01 cm? To 0.001 cm?

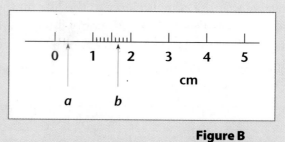

Figure B
For problem 6

†7. What fractions of a cubic centimeter do the smallest of the divisions on each graduated cylinder in Figure C represent? Express your answer in decimal form.

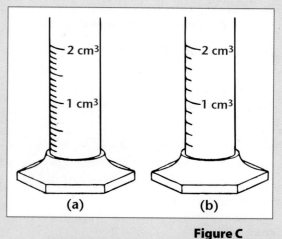

Figure C
For problem 7

8. A close look at Figure D shows that the top of the liquid in the graduated cylinder is not flat but curved. How do you decide how much water is in the cylinder?

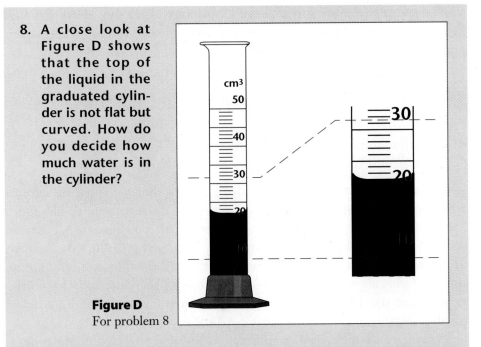

Figure D
For problem 8

9. Three students reported the length of a pencil to be 12 cm, 12.0 cm, and 12.00 cm. Do all three readings contain the same information?

10. What advantage is there to making graduated cylinders narrow and tall rather than short and wide?

EXPERIMENT
1.4 Measuring Volume by Displacement of Water

A granular solid like sand, although it does not flow as well as a liquid, can be measured by the same method. Suppose you have some sand in a cup. To find how much space the sand takes up in the cup, you could simply pour the sand into a graduated cylinder. But does the mark it comes to on the scale of the cylinder really show the volume of the sand? What about the air spaces between the loosely packed grains? The graduated cylinder really measures the combined volume of the sand plus the air spaces. However, you can do a simple experiment to find the volume of the sand alone.

Pour some sand into a dry graduated cylinder until it is about two-thirds full.

- What is the volume reading on the scale?

Now pour the sand into a beaker, and pour water into the graduated cylinder until it is about one-third full. Record the volume of the water, and then add the sand to the water.

- What is the volume of the sand plus the water?
- What is the volume of the sand alone?
- What is the volume of the air space(s) in the sand?
- What fraction of the dry sand is just air space? Express your answer in decimal form.

The experiment you have just done shows that we must be careful when we talk about the volume of a sample of a dry substance like sand. We must say how the volume was measured. If you have a bag of dry sand and want to know how many quart bottles it will fill, you need to know its volume dry. But if you want to know the volume of sand alone, not sand plus air space(s), then you must use a procedure like that in the experiment you have just done. You must measure the volume by liquid displacement.

Whenever we measure the volume of a solid by displacement of water, we make an assumption. We assume that the volumes of the solid alone and of the water alone add up to the volume of the solid and water together. This assumption may or may not be correct, depending on the kind of solid we have. For example, suppose you measure the volume of a few chunks of rock salt by the displacement of water. You will see that the total volume of rock salt and water becomes less as the salt dissolves. (See Figure 1.11.)

†11. A volume of 50 cm³ of dry sand is added to 30 cm³ of water for a total volume of 60 cm³.
   a. What is the volume of water that does *not* go into air spaces between the sand particles?
   b. What is the volume of water that does fill air spaces between the sand particles?
   c. What is the volume of the air spaces between the particles in the dry sand?
   d. What is the volume of the sand particles alone?
   e. What fraction of the total volume of the dry sand is sand particles? Express your answer in decimal form.

12. How would you measure the volume of granulated sugar?

13. How would you measure the volume of a cork stopper?

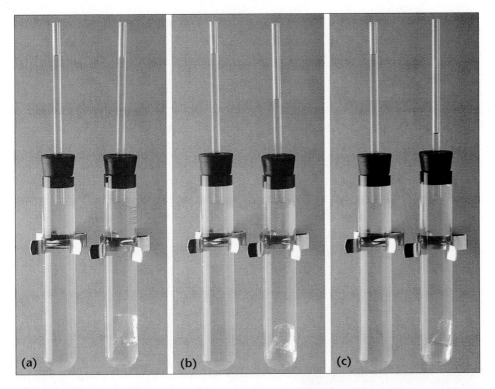

Figure 1.11

(a) A test tube containing only water, and another test tube containing water to which two pieces of rock salt have just been added. (b) and (c) The same test tubes 15 and 30 minutes later, after the salt has begun to dissolve. Notice the decrease in the total volume of the rock salt and water, as shown by the water level in the narrow glass tube. The test tube containing the salt was shaken several times to speed up the dissolving.

1.5 Shortcomings of Volume as a Measure of Matter

The experiment shown in Figure 1.11 strongly suggests that volume is not always a good measure of the amount of a substance. Here are some other difficulties with the use of volume to measure the amount of a substance. If you have ever pumped up a bicycle tire with air, you know that a gas is very compressible. As you push more and more gas into the tire, its volume remains almost unchanged. Does this mean that the amount of gas in the tire remains almost unchanged, too? If you compressed the gas obtained from heating baking soda by forcing it into a container of smaller volume, would there be less of the gas?

Finally, can we really use volume to compare the amounts of different substances, some of which are solids, some liquids, and others gases? Consider the heating of baking soda. Does measuring the volume of the baking soda, the liquid condensed near the top of the test tube, and the gas collected in the bottle really tell us how much of each of these substances we have?

1.6 Mass

The limitations of volume as a measure of the amount of matter must have been known to people many centuries ago. They developed a method for measuring the amounts of different substances independently of their volumes. From an Egyptian tomb several thousand years old, archaeologists have recovered a little balance arm of carved stone (Figure 1.12 (a)), with carefully made stone masses (Figure 1.12(b)). It was almost surely used, at the very dawn of history, for the careful measurement of gold dust. Goldsmiths knew even then that the balance was the best way to determine the amount of solid gold they had.

Figure 1.12 (a)
This balance, the earliest one known, comes from a prehistoric grave at Naqada, Egypt, and may be 7,000 years old. It uses limestone masses and has a red limestone beam 8.5 cm long and is shown in its true size. (*Courtesy of Science Museum, London*)

The balance was hung by the upper loop so that the horizontal bar was divided exactly into two arms of equal length. With no objects suspended from either arm, the balance bar would hang horizontally. An object hung from the loop on the end of one arm could be balanced by hanging some other objects from the end of the other arm.

No doubt, in using the balance, people soon learned that the bar would remain horizontal even though there were drastic changes in the shapes of the objects being balanced. Dividing a chunk of iron into a number of pieces or filing it into a pile of small grains does not affect the balance. A balance responds to something quite independent of the form of an object. What it responds to we call *mass*.

Suppose a piece of gold balances a piece of wood, and the piece of wood balances a piece of brass. Then we say that the masses of all three are equal. If something else balances the piece of brass, it also balances the wood and the gold and therefore has the same mass. The equal-arm balance gives us a way of comparing the masses of objects of any kind, regardless of shape, of color, or of what substance they are made.

Figure 1.12(b)
Some standard masses in units of beqa (BEK-ah) found in prehistoric graves in Egypt. The letters and numbers on these four masses were placed there by archaeologists. (*Courtesy of Science Museum, London*)

½ beqa

6 beqa

10 beqa

20 beqa

To record masses we shall need some standard masses with which various other pieces of matter can be compared. This standard mass is arbitrary—any mass, even the ancient Egyptian *beqa* may be chosen—but people must agree on it. In our work we shall use the *gram* (g), a unit of mass in the metric system. The international standard of mass in the metric system is a carefully made cylinder of platinum kept at Sèvres, near Paris, France, that has a mass of 1 *kilogram* (kg), or 1,000 g. All other kilogram masses are compared, directly or indirectly, with the standard whenever high precision is required. If you were to place a mass of 1 kilogram on a supermarket scale, the scale would read 2.2 pounds.

14. **When you buy things at a store, are they measured more often by volume or by mass? Give some examples.**

15. **What is your mass in kilograms?**

EXPERIMENT
1.7 The Equal-Arm Balance

The purpose of this experiment is to make you familiar with the basic operation of the simplest balance (Figure 1.13), and to help you to develop the necessary skills in using one. Whether or not most of your massing later in the course will be with an equal-arm balance, this balance provides a good starting point.

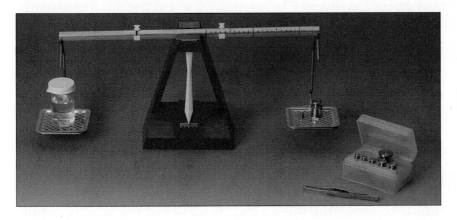

Figure 1.13
An equal-arm laboratory balance. The object to be massed is placed on the pan at the left. The standard gram masses are placed on the pan at the right. The tip of the pointer is at the middle of the scale on the base.

Make sure that the pans swing freely and that the vertical pointer in the center does not rub against the support. The pointer of the balance should swing very nearly the same distance on each side of the center of the scale when there is nothing on either pan. In order to adjust the balance so that it swings in this manner, first make sure that the pointed metal rider on the right arm is as near to the center of the balance as possible. Then move the rider on the left arm until the longer pointer on the center of the balance swings the same distance on each side of the center of the scale.

Your equal-arm balance comes with a set of masses, the smallest of which is 100 mg (1 mg, a *milligram*, is equal to 0.001 g. Thus 100 mg is equal to 0.100 g). Now that your balance is adjusted, use a set of gram masses to mass several objects of between 1 and 20 grams. (We shall abbreviate "to find the mass of" to the verb "to mass.")

Exchange objects with your classmates, and compare your measurements with theirs. Do not divulge your measurements until all students have recorded their measurements.

EXPERIMENT
1.8 Calibrating the Balance

Look carefully at several pennies. Do you think they all have the same mass? Would you expect them to differ a little in mass? Now mass the pennies on your balance. Record the mass of each penny in a table in your notebook, and be careful to keep track of which penny is which.

You have massed the pennies only to the nearest 0.1 g. How can they be massed to less than 0.1 g to see if there are tiny differences in their masses, smaller than 0.1 g? By using the rider on the right arm of the balance, you can measure masses to less than 0.1 g. Move the rider until it balances a 0.1-g mass placed on the left-hand pan, and then mark its position on the arm. Now make pencil marks on the arm, dividing into ten equal spaces the distance between the 0-g and the 0.1-g position of the rider. Each mark represents an interval of 0.01 g on this rider scale.

• How can you check to see if this is true?

If your balance has already been calibrated (that is, if there already is a scale marked on it), check to see if it is accurate.

1.9 Unequal-Arm Balances

When you calibrated your equal-arm balance, you placed a 0.1-g mass on the left-hand pan, and then moved out the rider on the right-hand arm until it balanced the 0.1-g mass. You marked the location of the rider, then divided the part of the arm between the rider and its zero position into ten parts. When you moved the rider to the 0.05 mark, it balanced a mass of 0.05 g. You learned that moving the rider farther in allowed it to balance a smaller mass.

Suppose we have a beam like the arms of the balance, with a pan on one end and a rider near the other. By trial, we can find a location where the beam can be balanced with the rider moved to a position near the supporting wedge, as shown in Figure 1.14. We mark this position of the rider "zero." Suppose we add some object of known mass to the pan. We find that we have to move the rider out to a new location in order to balance the beam again. We mark this new position. If we add a second object of equal mass to the pan, we must move the rider still farther out to balance the objects, and then we make another mark. We continue to do this until the rider can go no farther. We find our marks have been made at equal intervals, and we label them with the masses of the objects that we placed on the pan.

We can make a balance with two beams on the right, having riders of two different masses (Figure 1.15). The heavier rider lets us mass heavier objects. The lighter rider lets us make a more precise reading.

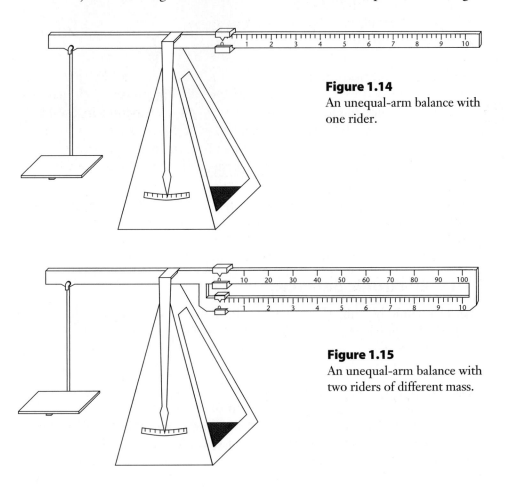

Figure 1.14
An unequal-arm balance with one rider.

Figure 1.15
An unequal-arm balance with two riders of different mass.

Figure 1.16
A common balance in a physician's office. (*Courtesy, Cardinal Scale Manufacturing Company*)

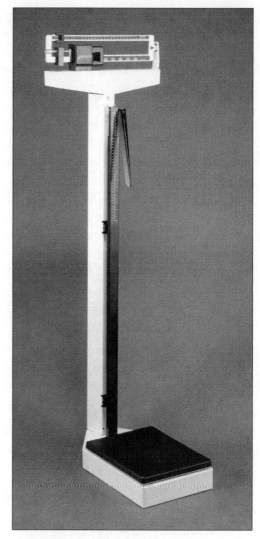

Such unequal-arm balances are very common. You probably have weighed yourself on one in your physician's office or in your school's health clinic (Figure 1.16). In such balances, the "beam" is not a single bar but is made of several parts. However, each rider is calibrated in the same way as the rider on the equal-arm balance.

You may very well be using an unequal-arm balance with four beams and four riders (Figure 1.17). Unlike the equal-arm balance that is pictured in Figure 1.13, unequal-arm balances are calibrated at the factory and their calibration marks cannot be changed.

Unequal-arm balances also differ from equal-arm balances in that they normally have only one pan. This single pan is used to hold the object being massed. There is no need for a mass set with an unequal-arm balance. Often these balances are called "single-pan balances."

1.10 Electronic Balances

An electronic balance (Figure 1.18) does not look at all like a balance. A balance compares the mass of the object on the pan with standard masses. The standard masses are placed either on a second pan or as riders on beams. An electronic balance has neither a second pan nor beams carrying riders.

If you press down gently on the pan of an electronic balance, you will notice that the pan moves very slightly. The change in the vertical position of the pan depends on the mass that is placed on it. This change in

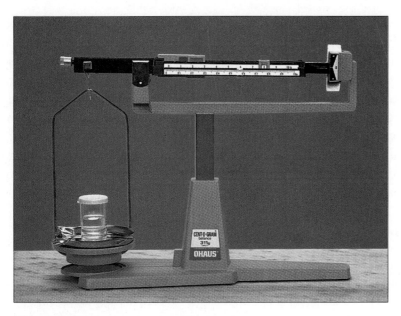

Figure 1.17

A single-pan balance with four beams and four riders. Unlike the equal-arm balance, this balance comes to rest rather quickly. This shortens the time required for massing.

position is translated by the electronic circuit into a number shown on the digital display. By first calibrating an electronic balance with known masses, we can then use it to measure unknown masses.

Figure 1.18

A general view of an electronic balance.

Electronic balances, or top loaders, as they are often called, have two advantages. First, it takes only seconds to mass an object. Thus, many teams in the laboratory can share one balance. Second, in many experiments you will need to find the mass of a liquid or powder in a container. A top loader can subtract the mass of the container and give you the mass of its contents directly. To use this feature, you press the *tare* button with the empty container on the pan. Then you mass the filled container.

Top loaders are delicate instruments and must be handled with care. In particular, use them only in the range for which they were designed. Your balance should probably not be loaded with more than 100 g.

EXPERIMENT
1.11 The Sensitivity of a Balance

If you mass the same object several times, will you find the same mass each time? And a related question: Will you detect the addition of small known masses on the balance, no matter how small they are? The purpose of this experiment is to find out how much you can trust your balance in answering such questions.

How you get the balance ready for the experiment will depend on the kind of balance available in your classroom. If you are working with an equal-arm balance, skip the next two paragraphs. If you are working with a single-pan balance, read the next paragraph but skip the following one. If you have a top loader, skip only the next paragraph.

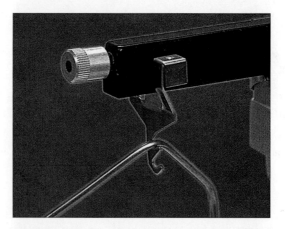

Figure 1.19
A close-up of a single-pan balance. The zero-adjustment knob is at the left end.

A single-pan balance, like the one pictured in Figure 1.17, must be "zeroed" before each use. To zero a balance means to adjust it to balance correctly under a single condition. In that condition nothing is on its pan and all riders are moved to the extreme left of their beams so they are at zero. The knob used to make this adjustment is shown at the upper left of the balance in Figure 1.19. Be sure to zero your balance while it is standing on a steady, level surface.

An electronic top-loading balance, like the one in Figure 1.18, should be placed on a steady, level surface and "zeroed" before it is used. To zero a balance means to adjust it to display "zero" when there is zero mass on its pan. The procedure for doing this is easily accomplished for top-loading balances, but differs from model to model. Your teacher will tell you how to zero the top-loading balance that you will be using.

Let us return to the first question: If you mass the same object several times, will you find the same mass each time? Mass two objects, such as a penny and a rubber stopper, alternately several times. To avoid being influenced by your previous results, have your lab partner read and record the results. Using two objects will make it necessary for you to change the settings. If you work with either an equal-arm balance or a single-pan balance, try to read the rider position as accurately as possible. Change roles with your lab partner, and then compare your results.

- Does your balance give reproducible results to the nearest 0.01 g? The nearest 0.001 g, or something in between?

Now for the second question: How much must the masses of two objects differ before your balance is able to distinguish between them? To obtain small, known masses, you will cut a group of smaller squares from a large square of graph paper that you have massed. Begin by neatly cutting a square of graph paper 20 squares along an edge. This large square is made up of 400 small squares.

Estimate the mass of the large square to the nearest 0.1 g. You can write your estimate on the square and compare the result with your classmates. Now mass the large square.

- What is the mass of your large square of graph paper?

Calculate the mass of a single small square. For the purpose of this experiment it will be convenient to have groups of squares with a mass between 0.003 and 0.007 g.

- How many of your small squares are in a group that has a mass between 0.003 and 0.007g?
- What is the mass of such a group of small squares?

  Now cut ten of these groups of small squares. With a penny or rubber stopper already balanced, add the groups of small squares one by one and observe the effect on the balance.

- How many groups of small squares did you have to add before the balance gave you a reliable response?

  The smallest change in mass that a balance can detect in a reproducible way is called the sensitivity of the balance.

- What is the *sensitivity* of your balance?

16. Aram masses an object on his equal-arm balance. By mistake he places the object in the pan on the same side as the rider. He balances the object by placing 4.500 g in the opposite pan and by setting the rider to 0.060 g. What is the mass of the object?

†17. Karen massed an object three times, using the same equal-arm balance and gram masses. Her results were: 18.324 g, 18.308 g, and 18.342 g. How could she best report the mass of the object?

18. Five students in turn used the same balance to measure the mass of a small dish. None knew what results the others obtained. The masses they found are given in the table below.

Student	Mass (g)
1	3.752
2	3.755
3	3.752
4	3.756
5	3.760

   a. Can you tell whether any student made an incorrect measurement?
   b. Do you think there is anything wrong with the balance?
   c. What do you think is the best way to report the mass of the dish?

19. Susan massed an object three times, using the same single-pan balance. Her results were: 21.420 g, 21.425 g, and 21.410 g. How could she best report the mass of the object?

20. Five students in turn used the same single-pan balance to measure the mass of a small dish. None knew what results the others obtained. The masses they found are given in the table below.

Student	Mass (g)
1	4.360
2	4.370
3	4.365
4	4.360
5	4.355

a. Can you tell whether any student made an incorrect measurement?

b. Do you think there is anything wrong with the balance?

c. What do you think is the best way to report the mass of the dish?

21. Five students in turn used the same electronic balance to measure the mass of a small dish. None knew the results of the others. The masses are given in the table below.

Student	Mass (g)
1	4.36
2	4.37
3	4.36
4	4.37
5	4.36

a. Can you tell whether any student made an incorrect measurement?

b. Do you think there is anything wrong with the balance?

c. What do you think is the best way to report the mass of the dish?

FOR REVIEW, APPLICATIONS, AND EXTENSIONS

22. Suppose the volume of a piece of glass is measured by displacement of water and by displacement of burner fuel. How would the two measurements compare?

23. In determining the volume of a rectangular box, five cubes were found to fit exactly along one edge, and four cubes to fit exactly along another edge. However, after six horizontal layers had been stacked in the box, a space at the top was left unfilled.

a. If the height of the space was half the length of an edge of a unit cube, what was the volume of the box?

b. If the height of the space was 0.23 of the length of an edge of a unit cube, what was the volume of the box?

24. What is the total number of cubes that will fit in the space enclosed by the dashed lines in Figure E? Is there more than one way to find an answer?

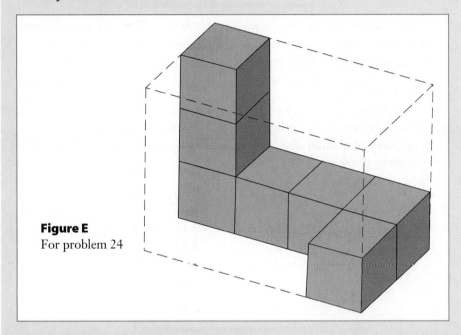

Figure E
For problem 24

25. In an experiment in which the volume of dry sand is measured by the displacement of water, the sand was slightly wet to begin with. What effect would this have on the volume of air space that was calculated? On the percentage of the volume that was air space?

26. a. How would you measure the volume of a sponge?

b. What have you actually measured by your method?

c. Does this differ from your measurement of the volume of sand?

27. Fuel oil usually is sold by the gallon, gas for cooking by the cubic foot, and coal by the ton. What are the advantages of selling the first two by volume and the last by mass?

28. In the following list of ingredients for a recipe, which are measured by volume, which by mass, and which by other means?

$1\frac{1}{2}$ pounds ground chuck   pinch of pepper
1 medium-size onion   3 drops steak sauce
$\frac{1}{2}$ cup chopped green pepper   oregano to taste
4 slices day-old bread   3 tablespoons oil
1 teaspoon salt   1 1-pound can tomato sauce

29. a. What is the volume of an aluminum cube with edges that are 10 cm long?
    b. What is the mass of this aluminum cube? (One cubic centimeter of aluminum has a mass of 2.7 g.)

30. One cubic centimeter of gold has a mass of 19 g.
    a. What is the mass of a gold bar 1.0 cm × 2.0 cm × 25 cm?
    b. How many of these bars could you carry?

31. Suppose that you took home an equal-arm balance. When you were ready to use it, you found that you had forgotten a set of gram masses.
    a. How could you make a set of uniform masses from materials likely to be found in your home?
    b. How could you relate your unit of mass to a gram?

32. Suppose you wanted to find the mass of water in a plastic bottle, and you took the following measurements using your equal-arm balance.

Mass of bottle and water	21.48 g
Mass of empty bottle	9.56 g
Mass of water	11.92 g

After you completed your measurements and calculations, you saw that you forgot to set the left-hand rider correctly; the beam was not level when the right-hand rider was on the zero mark and nothing was on the pans of the balance. Must you repeat the measurements to obtain the mass of the water?

33. Suppose you lost the rider for your scale. Try to think of another method, not using a rider, by which you could measure hundredths of a gram.

34. Suppose you balance a piece of modeling clay on the balance. Then you reshape it. Will it still balance? If you shape it into a hollow sphere, will it still balance?

35. Estimate in grams the mass of a watch. Now find the mass of a nickel (5¢) on your balance. Estimate the mass of the watch again. Did you change your estimate? Does knowing the mass of a nickel help you to better estimate your own mass? Why?

36. How could a person easily tell which letters had four and which five pages without opening the envelopes? What assumption did you make in arriving at your answer?

THEMES FOR SHORT ESSAYS

1. Suppose you are employed as a technical writer by a company that manufactures graduated cylinders. Printed instructions are to be included in packages sent out by the company. Write instructions telling customers how to use the cylinders correctly to measure the volumes of liquids.

2. A friend wants to use your balance during the summer. Write a complete set of instructions for her so that she will be able to do so successfully on her own without anybody being present to help her.

Chapter 2
Mass Changes in Closed Systems

EXPERIMENT
2.1 The Mass of Dissolved Salt

In Section 1.5, you learned that as salt dissolves in water, the combined volume of salt plus water decreases. This leads us to ask whether the mass also decreases when salt is dissolved in water.

Pour about 2 g of salt into the cap of a small plastic bottle, and carefully put it aside. Pour water into the plastic bottle until it is about two-thirds full. Find the total mass of the bottle, water, cap, and salt when all are on the balance together but the salt and water are not mixed.

Carefully pour the salt into the bottle, and put on the cap. Shake the bottle occasionally to speed up the dissolving of the salt. After the salt has dissolved, mass the capped bottle.

- Taking into consideration the sensitivity of the balance, what do you conclude about the mass of salt and water as the salt dissolves?

Answering this question calls for a comparison between two mass measurements, before and after dissolving. A good way to do that is to subtract the earlier value from the later value. This difference is called the *change* in mass. Suppose that the earlier mass was 28.36 g and the later mass was 28.37 g. Then the change is

$$28.37 \text{ g} - 28.36 \text{ g} = 0.01 \text{ g}.$$

In this example the later mass was greater than the earlier one. Suppose the earlier mass were 26.12 g and the later mass were 26.10 g. The change would still be calculated by subtracting the earlier value from the later value:

$$26.10 \text{ g} - 26.12 \text{ g} = -0.02 \text{ g}.$$

Thus, change is expressed by a signed number; it can be positive, zero, or negative.

Your measurements sum up the result of a single experiment. To have more confidence in your conclusion, you would have to repeat the experiment several times to be sure that you have not spilled any salt or made an error in reading a scale. Many repetitions would use up much time and would be boring. So, instead of asking you to repeat this experiment, we shall bring together the results of all the experiments done by the whole class. We shall follow the same procedure with other experiments.

The first step is to display the data from all stations on the chalkboard. For future reference, copy that table in your notebook.

- Considering the sensitivity of the balance, and the results of the entire class, what does your class conclude about the mass of salt and water as the salt dissolves?

- Can you suggest any reasons why all the members of the class do not find the same change in mass?

†1. In Experiment 2.1, The Mass of Dissolved Salt, how could you recover the dissolved salt? How do you think its mass would compare with the mass of dry salt you started with?

†2. If the change in mass in Experiment 2.1, The Mass of Dissolved Salt, were –0.0001 g, would you have observed this change using your balance?

3. a. In daily language we express the sign (plus or minus) of a change by using different words. For example, Elizabeth gained 1 kg, Tom lost 2 kg. Express these statements in terms of change in mass.

   b. In the morning the temperature rose from 10°C to 14°C. In the afternoon the temperature fell from 16°C to 11°C. What were the changes in temperature in the morning and in the afternoon?

2.2 Histograms

You probably have heard the proverb "a picture is worth a thousand words." Replace "words" by "numbers" and you have a statement that is just as valid. To see for yourself, consider the data in Table 2.1. They were taken by a large class.

Table 2.1 Class Data from Experiment 2.1, The Mass of Dissolved Salt

Team number	Mass before (g)	Mass after (g)	Change in mass (g)	Team number	Mass before (g)	Mass after (g)	Change in mass (g)
1	32.17	32.24	+ 0.07	11	31.31	31.26	– 0.05
2	29.57	29.36	– 0.21	12	32.55	32.38	– 0.17
3	28.67	28.67	0	13	32.26	32.21	– 0.05
4	38.02	38.02	0	14	34.29	34.27	– 0.02
5	27.67	27.64	– 0.03	15	31.43	31.39	– 0.04
6	31.55	31.55	0	16	31.86	31.86	0
7	31.45	31.43	– 0.02	17	30.86	30.87	+ 0.01
8	35.12	35.12	0	18	30.29	30.29	0
9	31.45	31.45	0	19	29.14	29.15	+ 0.01
10	28.67	28.65	– 0.02	20	30.86	30.64	– 0.22

All the data are in the table, but it is hard to grasp the content of the table all at once. Are the results of teams 1 and 2 typical of the entire class? Are the results of other teams close to either of them? Are the results of other teams somewhere in between the results of teams 1 and 2?

A glance at the "Change in mass" columns will show that the value of the change in mass that occurs most often is zero. There are also a few entries of +0.01 g and –0.02 g. Are they significantly different from zero? The change in mass was calculated by subtracting the mass before dissolving from the mass after dissolving. Each of these measurements has an uncertainty of ±0.01 g. For example, the results of team 3 could have been 28.67 g and 28.66 g, yielding a change of –0.01 g. To be able to draw conclusions from the data, we need to present them in a way that will emphasize an overall trend, with less detail.

Here is a simple way of doing just that. Imagine we arrange a number of bins or boxes marked "–0.03 to –0.01," "–0.01 to +0.01," and so on. Each team in the class writes its result for the change in mass on a card and deposits the card in the bin labeled with the appropriate interval (Figure 2.1). What should be done with values that are on the boundary between two bins? The simplest solution is to put them always in the bin to the right, as was done in Figure 2.1.

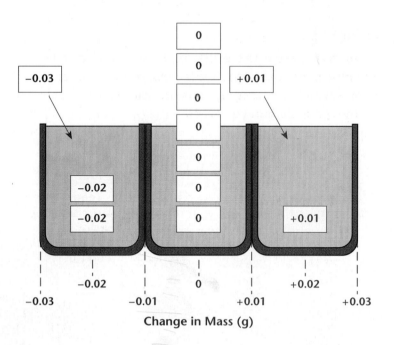

Figure 2.1
"Bins" for depositing data. The numbers in the upper row show the centers of the intervals represented by the bins. The ones in the lower row show the boundaries.

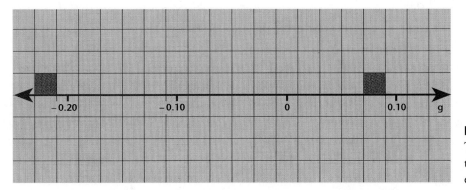

Figure 2.2
The starting points of a histogram: fitting the data on both ends of the horizontal axis.

Once all the cards have been placed in the bins, we count the cards in each bin and plot a bar graph of the results. Of course, we do not need real bins. We can use squares on graph paper instead. As with all graphs, it is a good idea to find first the largest and smallest values in order to know how to mark the scale on the horizontal axis (Figure 2.2).

Now we can go through Table 2.1 and mark a square for each value in the table. If a value has already appeared, we draw its square on top of the one below it. The final result is shown in Figure 2.3. The height of each column tells us the number of measurements that belong in each bin. A plot like Figure 2.3 that presents data by the number of times a value appears in an interval is called a *histogram*.

A histogram enables us to make several immediate observations. Most of the data are bunched around zero, with zero being the most frequent result. Some regions on both sides of zero have no data at all. But of the few readings far from zero, there are more on the left, indicating a loss, than on the right, indicating a gain.

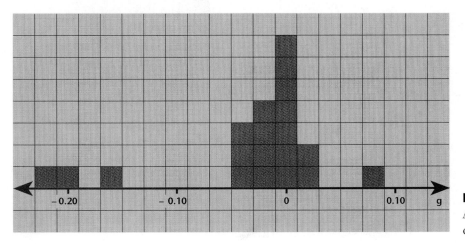

Figure 2.3
A histogram of the data in Table 2.1.

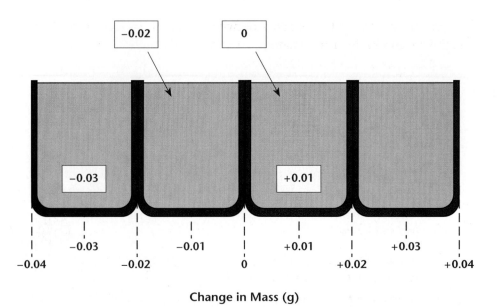
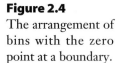

Figure 2.4
The arrangement of
bins with the zero
point at a boundary.

Change in Mass (g)

How would the data of Table 2.1 appear had we placed the zero
point at the border between two intervals (Figure 2.4)? Following the
convention that a border-value is placed in the interval to the right, the
resulting histogram is shown in Figure 2.5. The histograms in Figures
2.3 and 2.5 have the same general appearance. However, the histogram
in Figure 2.5 seems to suggest that the most frequent result is a small
gain in mass. This apparent conclusion is the result of having chosen
the zero point at the border between intervals, thereby shifting many
values to the right. Thus, to avoid drawing erroneous conclusions from

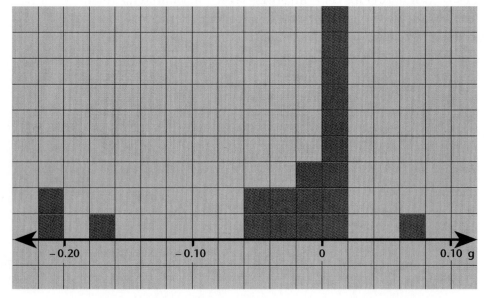

Figure 2.5
A complete histogram
of the data in Table 2.1,
with an interval width of
0.02 g and with zero at a
boundary.

histograms, try to set the borders between intervals so that frequent values fit into intervals rather than on their borders.

4. a. What is the interval immediately to the left of the interval containing zero in Figure 2.1?
   b. Identify the teams whose data appear in this column.

5. a. What is the end of the interval immediately to the right of zero in Figure 2.5?
   b. Identify the teams whose data appear in this column.
   c. Identify the teams whose data are in the last column on the left in Figure 2.5.

6. a. Draw a histogram of the data in Table 2.1. Use intervals with a width of 0.01 g and place zero at the center of an interval.
   b. If you drew the histogram correctly, the interval immediately to the left of zero should be empty. Why?
   c. This observation is lost in Figure 2.3. Therefore, is the new histogram better than the one shown in Figure 2.3? Explain your answer.

2.3 Using a Computer to Draw Histograms

Drawing a histogram can be divided into two main steps: (1) deciding the width of the bins and the value of the center of one bin, and (2) placing the values in the bins and counting them. The first step requires judgment, the second step is quite mechanical. Therefore, making the decisions is best done by you, but the counting and plotting can be left to a computer.

Figure 2.6 shows a histogram of the data in Table 2.1 created by a computer using a program written for *IPS*. Notice that the number of entries in each column must be read on the vertical axis on the left. Having the computer do the counting and drawing is to your advantage. You can try different histograms in a short time and find the one that in your opinion best summarizes the data.

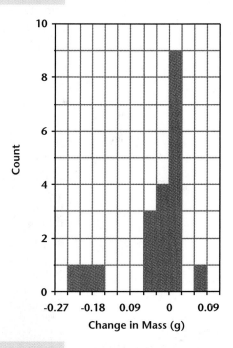

Figure 2.6

A computer-generated histogram of the data in Table 2.1.

7. a. What is the width of an interval in the histogram shown in Figure 2.6?
   b. Is zero at the center or at the border of the interval?

8. **Figures A and B also show histograms made from the data in Table 2.1. In each of these histograms,**
   a. **what is the width of the intervals?**
   b. **is zero at the center of an interval or at a border?**

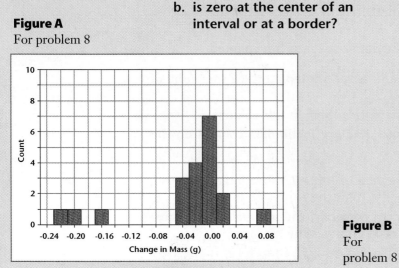

Figure A
For problem 8

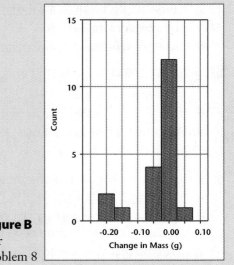

Figure B
For problem 8

EXPERIMENT
2.4 The Mass of Ice and Water

Here is another process involving a volume change. When ice melts, it contracts—its volume decreases. Does its mass also change?

Mass a small container with its cover; then put in an ice cube and mass again.

- What is the mass of the ice?

After all the ice has melted (if the container is not transparent, you can tell by shaking it), mass again.

- Do you notice any condensation on the outside of the container?
- If so, what should you do about it?
- From a histogram of class data, what do you conclude about change in mass when ice melts?

EXPERIMENT
2.5 The Mass of Copper and Sulfur

The changes you have examined so far were quite mild. A more dramatic change in matter takes place when sulfur and copper are heated together. Does the total mass change when these substances are heated together?

Put about 2 g of granular copper and about 1 g of sulfur in a test tube.

CAUTION: Do *not* use copper powder or copper dust. Always wear safety glasses when you use a burner.

Mix with a glass rod or shake gently. Close the end with a piece of rubber sheet held in place by a rubber band. Record the total mass of the closed tube. Heat the mixture gently until it begins to glow; then remove the flame immediately.

CAUTION: Do *not* touch the test tube until it has cooled.

• Has the total mass of copper and sulfur changed?

Describe the appearance of the material in the test tube.

• Do you think the substance in the bottom is sulfur, copper, or a new substance?

†9. The following data were recorded in an experiment in which copper and sulfur were made to react.

	Mass (g)
Tube and cover	20.484
Tube, cover, copper, and sulfur before reaction	23.440
Tube, cover, and products after reaction	23.386

  a. What is the mass of the substances before the reaction?
  b. What is the apparent change in mass of the reacting substances?
  c. What is the apparent percentage change in mass of the reacting substances?

10. A test tube containing 4.00 g of iron and 2.40 g of sulfur was heated in a manner similar to that of the copper-and-sulfur experiment. Before heating, the total mass of the tube and contents measured on the balance was 36.50 g. After heating, the mass was measured again. The mass of the tube and contents was 36.48 g.

  a. Do you think it reasonable that mass remained the same—was conserved—during this experiment?
  b. What additional steps would you take to increase your confidence in your answer to part (a)?

EXPERIMENT
2.6 The Mass of a Gas

In this experiment, a solid and a liquid produce a gas. Is there a change in mass?

CAUTION: Be sure to use only the small thick-walled bottle wrapped with tape that your teacher has provided. Always wear safety glasses when you work with gases.

Fill the bottle one-third full of water, then mass the bottle, its cap, and *one-eighth* of an Alka-Seltzer tablet. Place the piece of tablet in the bottle. Immediately screw the cap on very tightly and put it back on the balance.

- Does what happens inside the bottle affect the mass of the bottle and its contents?

  Slowly loosen the cap.

- Do you hear gas escaping?

  Again mass the cap, the bottle, and its contents.

- What do you conclude?

2.7 The Conservation of Mass

What have the last four experiments shown? If you have worked carefully, you have found that all the changes in mass that you observed were within the experimental uncertainty of your equipment. Therefore, your results agree with the conclusion that there was no change in mass that you could measure. From these experiments alone, you cannot predict with confidence that there will be no change in mass under other circumstances. For example, if you use larger amounts of matter in the experiments and use a balance of higher sensitivity, you might measure a change greater than the range of experimental uncertainty. Then you would conclude that mass really does not remain the same. Furthermore, although you checked four rather different kinds of change, there is an endless variety of other reactions you could have tried, some even more violent than the reaction of copper and sulfur.

What would happen, for example, if you set off a small explosion inside a heavy steel case, making sure nothing escaped? The experiments you have done give no direct answer to this question. But we can make the guess that the results of these four experiments can be generalized in the following way: In all changes, mass is exactly conserved, provided nothing is added (like the water that condensed on the outside of the closed container in the experiment with ice and water) and nothing escapes (like the gas in the last experiment). An experimental setup in which these conditions are satisfied is called a *closed system*. The generalization we have just stated is known as *the law of conservation of mass*. It holds for closed systems. It has been checked to one part in a billion*

---

*A billion is 1,000,000,000. Such a number is clumsy to write. Most of the zeros can be dispensed with by writing the number as 10^9 and reading it "ten to the ninth." The 9 is called an exponent and tells how many times we multiply 1 by 10 to get the number. For example, $1 \times 10 \times 10 = 10^2$, $1 \times 10 \times 10 \times 10 = 10^3$, and so on. We shall use this way of expressing numbers, called powers-of-10 notation, whenever it is convenient.

for a large variety of changes. That is, experiments have been done in which a change in mass of one billionth of the total mass would have been observed if it had occurred.

Still, all this vast amount of evidence in favor of the law of conservation of mass does not prove that it will hold forever under all conditions. Surely, if someone claimed that he or she had done an experiment in which as much as one millionth of the mass disappeared or was created, we should treat the results with great suspicion. First of all we should make many checks to determine whether there had been a leak of some sort in the apparatus from which, say, gas could escape. The chances are that we would find such a leak. On the other hand, suppose an experiment were done in which a change in mass of one part in 100 billion was reported. We might have to conclude, after a thorough examination of the experiment, that the law of conservation of mass has its limitations, that it holds to one part in a billion but not to one part in 100 billion (10^{11}). To date, no exceptions to the law of conservation of mass have been observed.

We have seen in this chapter that volume is very often a convenient way of measuring the quantity of matter. But we have also found out that, when matter changes form (when ice melts, salt dissolves, and so on), there is often an easily measurable change in volume but no observable change in mass: mass is conserved. It is the conservation of mass that makes mass such a useful measure of matter.

11. You wish to find your dog's mass, but the dog does not want to stand on the platform of the bathroom scale. You take the dog in your arms and stand on the scale; the mass indicated is 63 kg. Then you stand alone on the scale; the mass indicated is 55 kg.
   a. What is the mass of your dog?
   b. Give your reasoning.

12. a. Express the following numbers in powers-of-10 notation.
      100   10,000   100,000,000
   b. Write the following numbers without using exponents.
      10^5 10^6 10^9
   c. Express the following numbers in powers-of-10 notation.
      1,000   5,280   93,000   690,000
   d. Write the following numbers without using exponents.
      5.0×10^3 10^7 1.07×10^2 4.95×10^4

2.8 Laws of Nature

The law of conservation of mass is the first of several laws of nature that we shall study in this course. It is worthwhile to pause at this point and compare the laws of nature with the laws of society. Laws of society are legislated; that is, they are agreed upon and then enforced. If evidence is presented that you have broken such a law, you are punished. The laws of society can also be changed or repealed.

Laws of nature are different. Some laws began as lucky guesses based on a few crude experiments. Others were generalizations based on the analysis of many sophisticated experiments. Regardless of how it comes about, a person takes a risk in announcing what he or she believes to be a law of nature. Such a statement must stand the criticism and evaluation of many people doing additional experiments before it is recognized as a law of nature.

If you do an experiment that appears to violate a law of nature, you will not be punished. On the contrary, if you present convincing evidence that the law is not quite true, the law is changed to take into account the new experience. Only rarely does this lead to a complete repeal of the law. In most cases the change is a recognition of the limitation of the law.

13. There is an old saying: "What goes up, must come down." Does this express a law of nature? Why, or why not?

FOR REVIEW, APPLICATIONS, AND EXTENSIONS

14. Suggest a reason for putting the lid on the small container that you used in studying the mass of ice and water.

15. In Experiment 2.4, The Mass of Ice and Water, would the mass of the container and its contents stay the same if you started with water and froze it? Try it.

THEME FOR A SHORT ESSAY

The terms "conserve" and "conservation" have different meanings in science and in everyday life. In a dry summer we are urged to conserve water. Driving at moderate speeds helps in the conservation of fuel. Write a brief essay explaining the difference in the meaning of "conservation" in this chapter and in daily life.

Chapter 3
Characteristic Properties

3.1 Properties of Substances and Properties of Objects

How do you know when two substances are different? It is easy enough to distinguish between wood, iron, and rock, or between water and milk; but there are other cases in which it is not so easy. Suppose that you are given two pieces of metal. Both look equally shiny and feel equally hard. Are they the same metal? Or think of samples of two liquids. Both liquids are transparent and have no smell. Are they the same or different?

To answer such questions, we shall have to do things to substances in order to reveal differences that are not directly apparent. Merely massing the two pieces of metal will not do. Two objects can be made of different materials and yet have the same mass. Think of a 100-g steel cylinder and a 100-g brass cylinder of the kind used as masses on a balance. On the other hand, two objects can have different masses and be made of the same material. For example, consider two hammers, both made of steel but one much larger and with a greater mass than the other. Mass is a property of an object. It is not a property of the substance of which the object is made.

To find out if two pieces of metal that look alike are made of the same substance, you may try to bend them. Again, one may be thick and hard to bend, and the other may be thin and easy to bend; yet they both may be made of the same substance. On the other hand, you may find that two pieces of metal of different thicknesses but made of different substances bend with equal ease. Thus, ease of bending is also a property of the object and not of the substance.

Suppose we want to find out whether two objects are made of the same substance or of different ones. We have to test for properties that are characteristic of substances and not properties such as mass, volume, or shape. The latter are properties of objects. In this chapter we shall concentrate on properties that show differences between substances. These we call *characteristic properties*.

---

1. State which words in the following descriptions refer to properties of the substances and which refer to properties of the objects.
   a. A sharp, heavy, shiny, stainless-steel knife
   b. A small chunk of black tar
   c. A beautifully carved wooden chair

EXPERIMENT
3.2 Freezing and Melting

If you live in a part of the country where it snows in the winter, you know that a big pile of snow takes longer to melt than a small one. Does this mean that the big pile melts at a higher temperature? Let us see whether the temperature at which a sample of a substance melts or freezes is really a characteristic property of the substance. To do so, we shall measure the freezing temperatures of some substances by using samples of different mass. For convenience, we shall use substances that freeze above room temperature.

Your teacher will distribute test tubes containing different amounts of a solid substance. Immerse the test tube in a water bath. Heat the water until the solid in the test tube is completely melted.

CAUTION: Always wear safety glasses when you use a burner.

Insert a thermometer into the molten substance. Make sure that the solid in the test tube is completely melted before you remove the burner. For comparison, it may be interesting also to measure the temperature of the water with a second thermometer. (See Figure 3.1.)

While the liquid in the test tube cools, measure and record both temperatures every half-minute. Stirring the liquid and the water with two thermometers will ensure that the temperatures will be the same throughout each substance.

In addition, record the temperature of the molten substance just as it begins to solidify. Continue to take readings every half-minute until the temperature of the substance drops to about 65°C.

You can stop the stirring in the test tube when the thermometer is no longer free to move.

CAUTION: To prevent breakage, do not remove the thermometer from the test tube.

Figure 3.1
Apparatus used to obtain data for the cooling curve of a liquid as it cools and freezes. The thermometer in the test tube measures the temperature of the liquid; the one in the beaker measures that of the water bath. Note that the level of the molten solid in the test tube is below the level of the water in the beaker.

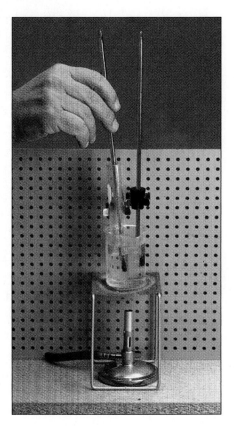

- In your table of temperatures and times, do you note any difference in the way the substance and the water cooled?

A better way to display your results is to plot the temperatures as a function of time on graph paper. Draw the graph of the temperatures of the substance and the water using the same axes. Compare your graphs with those of your classmates.

- Do all graphs have a flat section?
- Does the temperature of the flat section depend on the mass of the cooling materials?
- Do you think that all the samples used in the class were of the same material?

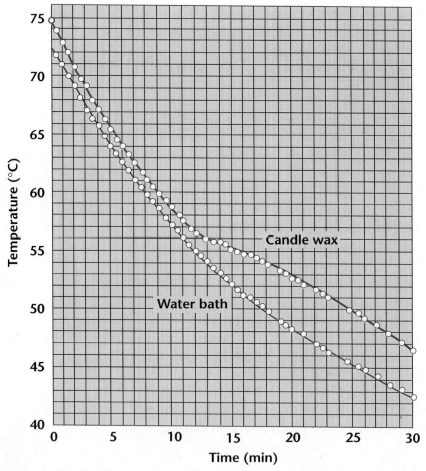

Figure 3.2
The cooling curves of candle wax and of the water bath surrounding the test tube holding the wax. The absence of a flat section in the curve of the wax means that candle wax has no freezing point.

You have now determined the freezing point of a substance by noting the plateau (flat section) in the cooling curve. For some substances the plateau is more easily recognizable than it is for others. Some cooling curves, however, may not have a flat section at all. For example, look at the cooling curve of candle wax shown in Figure 3.2. The data for this curve were obtained in the same way as in your experiment. The fact that no part of the curve is flat means that candle wax has no freezing point. In other words, there is no temperature at which it changes from liquid to hard solid without continuing to cool down during the process. Similarly, as you warm a piece of candle wax in your hand, it becomes softer and softer. However, there is no temperature at which it changes from hard solid to liquid without continuing to warm up.

It is harder to measure the melting point of a substance than to measure the freezing point. Since we cannot stir a solid, it is necessary to heat it very slowly and evenly until it melts. If we do so, we find that we get a curve with the flat portion at exactly the freezing temperature. A solid melts at the same temperature at which the liquid form of the same substance freezes.

†2. The graph in Figure A represents data from an experiment on the cooling of TOP. During which time intervals is there only liquid? only solid? both liquid and solid?

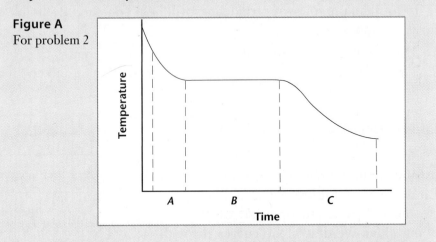

Figure A
For problem 2

3. Water freezes at 0°C. Sketch a graph of temperature as a function of time for a container of water at 20°C, after it is placed in a freezer at –10°C. Show the temperatures +20°C, 0°C, and –10°C on the vertical axis. Continue the graph until practically no further change will take place in the temperature of the ice.

3.3 Graphing

Drawing graphs effectively is a useful skill. Here are a few guidelines, using Figure 3.2 as an illustration.

The first step in drawing a graph is to set the scales for the horizontal and vertical axes. Let us begin with the horizontal axis. The table of data from which the graphs were made is not given in the text; but we can be quite sure that readings were taken over a period of 30 minutes. Thus, the time axis must cover 30 minutes. If it had covered only 20 minutes, many data points would have been lost. Note that there are 30 divisions on the horizontal axis, so each division corresponds to one minute. This scale makes it easy to locate whole minutes and half-minutes. Looking at the curves, you can recognize which readings were taken at whole minutes and which at half-minutes.

Now look at the vertical axis. The highest temperature reading was close to 75°C, and the lowest reading was close to 42°C. This is a span of 33 C°. Why did the scale not start at 42°C and end at 75°C? Consider the large rectangle to be a sheet of graph paper. It has 36 divisions along the vertical axis. If we had labeled the lower end of the axis as 42°C and the upper end as 75°C, there would have been 33°C spread over 36 divisions, that is, 33/36 or 0.92 C° per division. It would be very awkward both to plot points and to read their values on such a scale (Figure 3.3). Just try to mark 72.0°C on this scale!

To sum up, the following guidelines will help you choose suitable axes for your graphs:

1. The axes must cover the entire range of your data points.
2. There is no need to start the scale at the smallest value and end it at the largest value.
3. Be sure that each division corresponds to a unit that makes plotting and reading easy.

Once all the points have been plotted, we usually connect them with a smooth curve. Look again at Figure 3.2. Neither of the two curves passes through all the plotted points. Had we drawn the curves through all points, the curves would be full of little

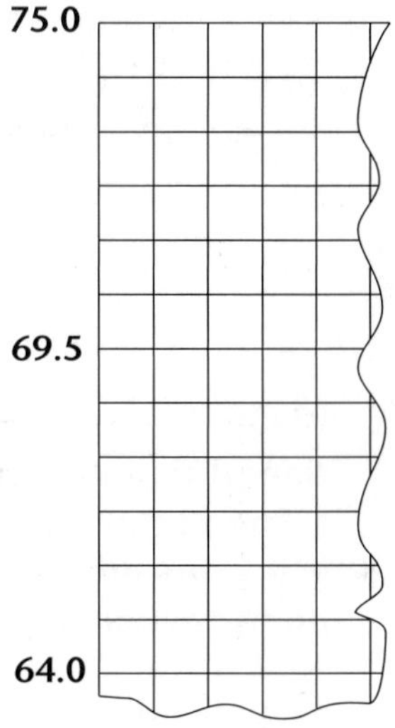

Figure 3.3
The upper left-hand corner of the graph paper in Figure 3.2 with the vertical scale made to extend from 42 to 75 over the 36 divisions.

wiggles. Because the points stand for measured values, they are subject to uncertainties. In all likelihood the small wiggles are the result of these uncertainties. Therefore, a smooth curve that is close to most points and shows the general trend is a better picture of the results than a curve that goes through all the points. In some cases a curve may not go through any data points.

4. a. What values correspond to points *A* through *D* on scale I of Figure B, and what values correspond to points *E* and *F* on scale II?

   b. Which points are easier to read and why?

Figure B
For problem 4

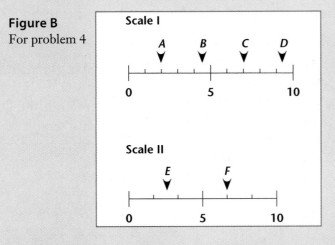

5. On which of the scales shown in Figure C would you choose to plot each of the following sets of numbers? In each case explain the reason for your choice.

   a. 9.0, 17.5, 29.5, 36.0

   b. 17, 23, 55, 62

   c. 23.5, 31.2, 34.8, 39.0

Figure C
For problem 5

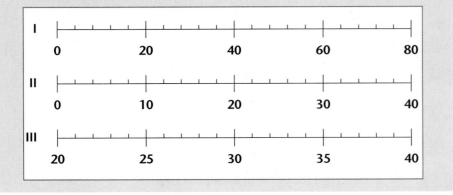

EXPERIMENT
3.4 Boiling Point

Everybody knows that it takes longer to bring a full pan of water to a boil than a half-filled one. Does this mean that the full pan gets hotter? To see what happens, heat either 10 cm³ or 20 cm³ of a liquid in a test tube in the apparatus shown in Figure 3.4. To prevent uneven boiling, add a few small chips of porcelain to the liquid. The glass and the rubber tubes will prevent vapors from spreading into the room. The vapors will condense in the cooled test tube.

CAUTION: Always wear safety glasses when you use a burner.

Some of the liquids you will be using may be flammable. To guarantee gentle heating, use the burner stand as shown in Figure 3.4.

Read the temperature of the liquid every half-minute until the liquid has been boiling for about five minutes. Then plot a graph of the temperature of the liquid as a function of time. Compare your results with those of other students in your class.

- Do all the graphs look alike at the beginning?
- Do all the graphs have a flat section?
- Was the temperature the same in all test tubes once the liquid began boiling?
- What does a difference in boiling point reveal?
- Does the boiling point of a liquid depend on the amount of liquid? Is boiling point a characteristic property?

Figure 3.4
A thermometer supported by a rubber stopper in a test tube measures the temperature of a liquid as it is heated to its boiling point. To prevent uneven boiling, a few small chips of porcelain are placed in the liquid. The water in the container should be cold.

3.5 Boiling Point and Air Pressure

Suppose your class measured the boiling points of some liquids very carefully and compared the results with those of *IPS* students in schools across the country. You would find that the other schools' results differed slightly from yours. The differences would be greater if the other schools were at different elevations. The boiling point of a liquid depends on the air pressure, which in turn depends on elevation and the weather.

You have changed air pressure many times. Think of a straw in a glass of water. The level of the water inside the straw is the same as the level outside because the air pressure is the same. When you drink through a straw, you remove some of the air, reducing the air pressure in the straw. The difference in pressure causes the water to rise inside the straw until the pressure is again equal.

How high can water rise inside a straw or some other tube? With a long enough tube and a pump to remove the air, the water will rise to a height of about 10 m at sea level. If we replace the water with mercury, the mercury will rise to 0.760 m or 760 mm. This is a much more convenient height to measure (Figure 3.5).

When the top of an evacuated tube containing mercury is sealed and a metric scale is placed next to it, we have a *barometer* (Figure 3.6). The

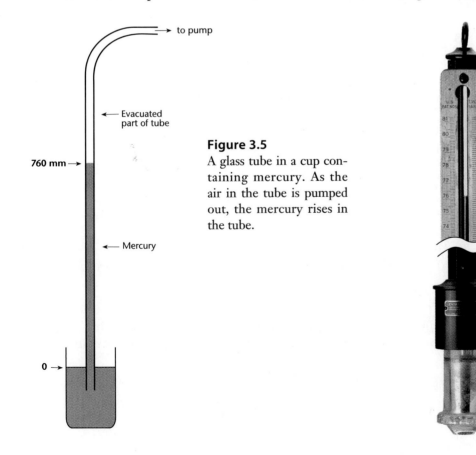

Figure 3.5
A glass tube in a cup containing mercury. As the air in the tube is pumped out, the mercury rises in the tube.

Figure 3.6
The upper and lower parts of a mercury barometer. The long middle section is not shown. The scale on the upper part is in both millimeters and inches. Note that the mercury container at the bottom is much wider than the long tube.

height of the mercury column will show variations. A higher column indicates that the air pushes harder on the surface of the mercury. The height of the column does not depend on the diameter of the tube. This is why the height of the mercury column is used as a measure for the air pressure generated by the atmosphere. This pressure is called *barometric pressure*. In weather reports barometric pressure is often expressed in inches of mercury.

When the barometric pressure is higher, the boiling point of a liquid is higher; when the pressure is lower, the boiling point is lower. The variations in barometric pressure at any one location are rather small. The highest and lowest pressures recorded in Boston, Massachusetts, during one year were 780 and 730 mm, respectively. At these extreme conditions, the boiling point of water varies by only about ±1°C.

The differences in barometric pressure between places at different elevations can be considerable, resulting in significantly different boiling points. The boiling points of water at different elevations are shown in Figure 3.7.

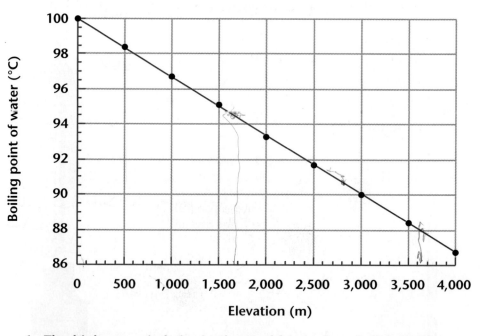

6. **The highest capital city in the world is La Paz, Bolivia, with an elevation of about 3,600 m. What is the boiling point of water in**

Figure 3.7
A graph of the boiling point of water as a function of elevation above sea level.

6. The highest capital city in the world is La Paz, Bolivia, with an elevation of about 3,600 m. What is the boiling point of water in La Paz?

7. The elevations of Memphis, Tennessee; Boulder, Colorado; and Los Alamos, New Mexico, are 80 m, 1,630 m, and 2,280 m, respectively. What is the boiling point of water in each of these cities?

EXPERIMENT
3.6 Mass and Volume

By now you have measured the mass and volume of different objects. Is there a relation between these two quantities?

The three long, thin cylinders shown in Figure 3.8 were cut from a uniform rod of aluminum. Using a 10-mL graduated cylinder, you can verify that each of them has a volume of 1.0 cm³.

• Do you expect the mass of these three cylinders to be the same? Why or why not? Mass the three cylinders to check your expectation.

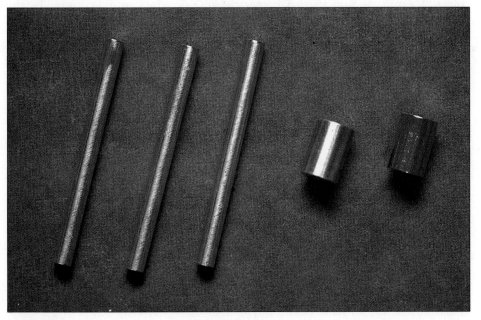

Figure 3.8
Five cylinders. Each has a volume of 1.0 cm³.

Suppose that the aluminum rod from which the three cylinders were made was cut into cylinders twice as long.

- What would be the mass of each cylinder?
- What would be its volume?
- How can you find the mass of one cubic centimeter of such a cylinder without cutting it in half?

The three cylinders with which you have been working have the same shape. Does the mass of a piece of aluminum depend on shape or only on volume? To answer this question, you will use another aluminum cylinder. Its volume is also 1.0 cm^3 and it fits into the 10-mL graduated cylinder.

- What is the mass of the short aluminum cylinder?

A piece of aluminum and a piece of brass can be cut so they have the same mass or the same volume. Can you make a piece of brass that will have both the same mass and the same volume as a piece of aluminum? To help answer this question, you can use a 1.0-cm^3 brass cylinder.

- What do you find?

3.7 Density

Rarely do we encounter samples of matter having exactly one unit volume (1 cm^3), as you did in Experiment 3.6, Mass and Volume. In the laboratory we usually work with samples that have a volume either larger or smaller than one unit volume. Nevertheless, we can still find the mass of a unit volume of a substance. We begin by measuring the mass and volume of any size sample of a substance. The mass of one unit volume can then be calculated from these measurements by dividing the mass of the sample by its volume.

Suppose a sample of aluminum was measured and found to have a volume of 4.8 cm^3 and a mass of 13 g. We calculate the mass of a unit volume of aluminum in the following way:

$$\frac{13 \text{ g}}{4.8 \text{ cm}^3} = 2.7 \text{ g/cm}^3.$$

We calculated the mass of a unit volume by dividing mass by volume. Therefore we refer to the mass of a unit volume as "mass per unit volume." The word "per" means a division by the quantity that follows it. For example, the speed of a car is stated in miles per hour—that is, a distance measurement divided by a time measurement. The mass per unit volume of a substance is called the *density* of that substance. Its units are g/cm^3 (grams per cubic centimeter). In the example shown above, the density of the aluminum is 2.7 g/cm³. The two statements, "The mass of 1 cm³ of aluminum is 2.7 g," and "The density of aluminum is 2.7 g/cm³" contain the same information. The second statement is shorter and is used more often.

8. A parking lot is filled with automobiles.
   a. Does the number of wheels in the lot depend upon the number of automobiles?
   b. Does the number of wheels per automobile depend upon the number of automobiles?
   c. Is the number of wheels per automobile a characteristic property of automobiles that distinguishes them from other vehicles?

9. a. Draw a graph of the number of wheels in a parking lot as a function of the number of cars.
   b. Draw a graph of the number of wheels per car as a function of the number of cars in the lot.

3.8 Dividing and Multiplying Measured Numbers

When you find the density of a substance from the measured mass and volume of a sample, your calculation uses numbers of limited accuracy. For example, consider a pebble that has a mass of 12.36 g and a volume of 4.7 cm³. Here the mass is given to four digits and the volume to two. If you calculate the quotient 12.36 g/4.7 cm³ on a calculator, you get the number 2.62978723. However, not even a calculator can produce numbers that are more accurate than the data used in the calculations.

For division and multiplication, it is good to remember a simple rule of thumb: The result should have as many digits as the measured number with the fewest digits. It is always advisable to calculate one additional digit, and then round off. In the example we just saw, the density is

$$\frac{12.36 \text{ g}}{4.7 \text{ cm}^3} = 2.6 \text{ g/cm}^3.$$

The only *significant digits* are the "2" and the "6." The remaining digits in the display of the calculator are not significant and should be dropped.

If we calculate the volume of an object from its dimensions, the same rule applies. Suppose the measured dimensions of a rectangular solid are

4.82 cm, 11.05 cm, and 1.28 cm. Then the volume of the solid is correctly reported as

$$4.82 \text{ cm} \times 11.05 \text{ cm} \times 1.28 \text{ cm} = 68.2 \text{ cm}^3.$$

The volume should *not* be reported as 68.174080 cm³.

How do we count the digits in a decimal measurement whose last digit is a zero, as in the measurement 4.20 cm? (Section 1.3 explained why the zero is written.) In these cases, the zero is counted as a significant digit. However, in such measured numbers as 0.86 cm and 0.045 g, the zeros are not counted. Their only purpose is to locate the decimal point. Both these numbers are given to two significant digits.

10. **Do the following calculations to the proper number of digits.**
    a. 125
       23.7
    b. 20.5
       51.0
    c. 0.065
       32.5
    d. 1.23
       0.72

    e. 4.72 × 0.52
    f. 6.3 × 10.08

    g. 1.55 × 2.61 × 5.3
    h. 3.01 × 5.00 × 25.62

11. **Suppose the measurement of the first dimension of the rectangular solid discussed above were 4.81 cm or 4.83 cm instead of 4.82 cm. Write all the digits in the number that would represent the volume of the solid. Which of these digits are the same as those calculated in the example above?**

EXPERIMENT
3.9 The Density of Solids

Pick up two cubes that look alike and that have the same volume. Can you decide simply by handling them whether they have the same or different masses?

Mass the cubes on your balance.

• Which of the cubes has the greater density?

Now, just by handling them, compare the mass of each of the cubes with the mass of a third object, which has a different volume.

• Can you decide in this way if the third object is made of a different substance?

Measure the dimensions of each of the three objects as accurately as you can. Calculate the volume and then the density of each.

- Are you now able to decide whether or not the third object is made of the same substance as either of the other two cubes?

The volume of an irregularly shaped object is difficult to determine from a measurement of its dimensions. You can find its volume by the displacement of water, as described in Chapter 1. Use this method to find the density of an irregularly shaped stone. Compare the density of your stone with the results of other students who used pieces that came from the same rock.

- What possible reasons can you give for the different measured values of density?

12. What measurements and what calculations would you make to find the density of the wood in a rectangular block?

13. A student announced that she had made a sample of a new material that had a density of 0.85 g/cm³. Can you tell how large a sample she had made? How?

†14. A block of magnesium whose volume is 10.0 cm³ has a mass of 17.0 g. What is the density of magnesium?

15. Two cubes of the same size are made of iron and aluminum. How many times as heavy as the aluminum cube is the iron cube? (See Table 3.1, page 58.)

†16. a. A 10.0-cm³ block of silver has a mass of 105 g. What is the density of silver?

b. A 5.0-cm³ block of rock salt has a mass of 10.7 g. What is the density of rock salt?

c. A sample of alcohol amounting to 0.50 cm³ has a mass of 0.41 g. What is its density?

EXPERIMENT
3.10 The Density of Liquids

Examine two samples of liquid. Smell them and shake them, but don't taste them. Can you tell whether they are the same or different? Perhaps by finding their densities, you can answer the question. You can find the density of a liquid by massing it on a balance and measuring its volume with a graduated cylinder. The cylinder is too large to fit easily on the balance. You must mass the liquid in something smaller.

A small amount of liquid will cling to the inside of any container from which you pour it. Therefore, to be certain you mass the volume of the liquid you measure in the graduated cylinder, you must be careful of the order in which you make your measurements of mass and volume.

- Which is more accurate, to mass the liquid in the small container before pouring it into the graduated cylinder, or to determine its volume first?

- What are the densities of the two liquids?

- Are the two liquids the same or different?

17. Suppose you are given two clear, colorless liquids. You measure the densities of these liquids to see whether they are the same substance or different ones.
    a. What would you conclude if you found the densities to be 0.93 g/cm³ and 0.79 g/cm³?
    b. What would you conclude if you found the density of each liquid to be 0.81 g/cm³?

EXPERIMENT
3.11 The Density of a Gas

It is more difficult to measure the density of a gas than that of a liquid or a solid. Gases are hard to handle, and most of them cannot even be seen. In fact, early chemists neglected to take into account the mass of gases produced in experiments.

When we mix Alka-Seltzer tablets and water, a large volume of gas is produced. We can find the density of this gas by massing the tablets and the water before and after they are mixed and by collecting and measuring the volume of the gas. You will recall that, in Experiment 2.6, The Mass of a Gas, you measured the mass of some of this same gas, but you did not measure the volume.

Place two half-tablets of Alka-Seltzer and a test tube with about 5 cm³ of cold water on the pan of your balance as shown in Figure 3.9(a) or 3.9(b). Find the total mass of these objects.

Figure 3.9(a)

To support a test tube that contains water on the *IPS* balance, you can use a paper clip and a rubber band as shown. Be sure that the paper clip does not rub against the arm of the balance.

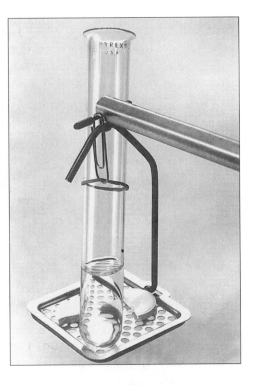

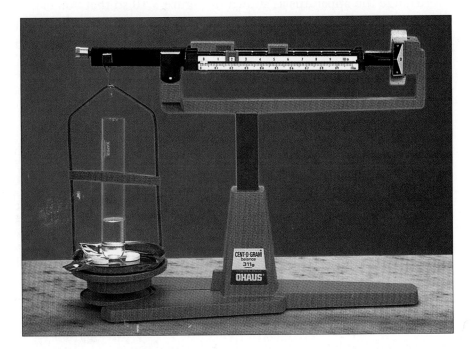

Figure 3.9(b)

A good way of supporting the test tube on a multibeam balance.

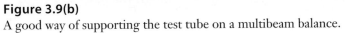

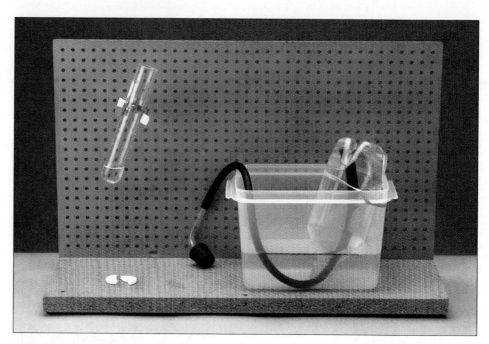

Figure 3.10
When the two half-tablets are added to the test tube, the gas generated is collected by displacing water from the inverted bottle on the right.

Arrange the apparatus as shown in Figure 3.10 so that you can collect the gas that will be produced. Be sure that the end of the rubber tube is at the top of the collecting bottle and that the whole length of the tube is clear and open.

CAUTION: Always wear safety glasses when you work with gases.

Drop the two half-tablets into the water, quickly insert the delivery tube and stopper into the test tube, and collect the gas produced. Practically all the gas will be produced in the first ten minutes of the reaction. At the end of this time, remove the delivery tube from the container and then remove the stopper from the test tube.

• Why is it important to hold your hand across the mouth of the bottle while removing it from the container?

Turn the bottle upright, and find the volume of the water that has been displaced by gas.

• How is this volume related to the volume of the gas?

• How can you find the mass of the gas?

- What is the density of the gas?
- What assumptions have you made in using this method?

†18. A mixture of two white solids is placed in a test tube, and the mass of the tube and its contents is found to be 33.66 g. The tube is stoppered, and apparatus is set up to collect any gas produced. When the tube is gently heated, a gas is given off, and its volume is found to be 470 cm^3. After the reaction, the mass of the test tube and its contents is found to be 33.16 g.
   a. What is the mass of the gas collected?
   b. What is the density of the gas collected?

19. Experiment 3.11, The Density of a Gas, is repeated with a sample of a different solid. Here are the data obtained:

Mass of solid, test tube, and water before action	35.40 g
Mass of test tube and contents after action	34.87 g
Volume of gas collected	480 cm^3

   Could this gas be the same as that which was produced in Experiment 3.11?

20. The volume of gas generated by treating 1.0 g of baking soda with 8.8 g of acid is 200 cm^3. The remaining acid and solid have a mass of 9.4 g. What is the density of the gas produced?

21. If the volume of gas in the preceding problem is compressed to 50 cm^3, what will be the density of the gas now? To what volume must the gas be compressed before it reaches a density of 1.0 g/cm^3, a typical density of a liquid?

22. The gas whose density you measured in Experiment 3.11 dissolves slightly in water.
   a. How does this affect the volume of the gas you collect?
   b. How does this affect your determination of the density of the gas?

3.12 The Range of Densities

Table 3.1 on the next page lists the densities of various substances. Note that most solids and liquids have a density that is between 0.5 g/cm^3 and about 20 g/cm^3. The densities of gases are only about 1/1,000 of the densities of solids and liquids.

Is the density of a substance the same under all conditions? Most substances expand when heated, but their mass remains the same.

Table 3.1 Densities of Some Solids, Liquids, and Gases (g/cm³)

Osmium	22.5	Oak	0.6–0.9	
Platinum	21.4	Lithium	0.53	
Gold	19.3	Liquid helium		
Mercury	13.6	(at –269° C)	0.15	
Lead	11.3	Liquid hydrogen		
Copper	8.9	(at –252°C)	0.07	
Iron	7.8	Carbon dioxide	1.8×10^{-3} *	
Iodine	4.9	Oxygen	1.3×10^{-3}	
Aluminum	2.7	Air	1.2×10^{-3}	at atmospheric
Carbon tetrachloride	1.60	Nitrogen	1.2×10^{-3}	pressure and room
Water	1.00	Helium	1.7×10^{-4}	temperature
Ice	0.92	Hydrogen	8.4×10^{-5}	
Methyl alcohol	0.79	Air (at 20 km altitude)	9×10^{-5}	

* Small numbers less than 1 can, like large numbers, be expressed in powers of 10. For example, we write 0.1 as 10^{-1}, 0.01 as 10^{-2}, 0.001 as 10^{-3}, and so on, using negative numbers as exponents.

If we have a decimal such as 0.002, we can write it first as 2×0.001, and then, in powers-of-10 notation, as 2×10^{-3}.

Another example: $0.00009 = 9 \times 0.00001 = 9 \times 10^{-5}$. The negative exponent of the 10 tells how many places the decimal point must be moved to the left to give the correct value in ordinary notation.

Therefore, the density depends on the temperature, becoming lower as the material expands and increases in volume. The expansion is very small for solids and liquids and has little effect on the density. The situation is quite different for gases, which expand greatly when heated. Moreover, solids and liquids are difficult to compress, but gases are easily compressed, as you probably know from pumping up a bicycle tire. Therefore, when measuring the density of a gas, we have to state the temperature and the pressure at which the density was measured.

23. Write each of the following numbers in powers-of-10 notation.
    a. 0.001   0.1   0.0000001
    b. 1/100   1/10,000

24. Write each of the following numbers as a number between 1 and 10, times the appropriate power of 10.
    a. 0.006   0.000032   0.00000104
    b. 6,000,000   63,700

25. Write each of the following numbers in ordinary notation.
    a. 10^{-2} 10^{-5} 3.7×10^{-4}
    b. 1.05×10^{-5} 3.71×10^{3}

26. A small beaker contains 50 cm^3 of liquid.
    a. If the liquid were methyl alcohol, what would be its mass?
    b. If the liquid were water, what would be its mass?

†27. The densities in grams per cubic centimeter of various substances are listed below. For each density indicate whether the substance is most likely to be a gas, liquid, or solid. (Refer to Table 3.1)
    a. 0.0015   b. 10.0   c. 0.7   d. 1.1   e. 10^{-4}

28. Estimate the mass of air in an otherwise empty room that is the size of your classroom.

3.13 Identifying Substances

We have looked for properties that can help us to distinguish between substances that appear to be the same. So far we have found three properties that do not depend on how much of a substance we have or on its shape. These properties are melting point, boiling point, and density.

Suppose you measured the melting points of two samples of matter and found them to be the same. If you then measured their boiling points and found that these were also the same, you might suspect that you had two samples of the same substance. You would not expect them to differ in their density or in any other properties. But, as Table 3.2 shows, we cannot depend on two properties alone to distinguish between substances, particularly if the measurements are not highly accurate.

Table 3.2 Some Substances with Similar Properties

	Density (g/cm^3)	Melting point (°C)	Boiling point (°C)
Group 1			
Methyl acetate	0.93	−98	57
Acetone	0.79	−95	57
Group 2			
Isopropanol	0.79	−89	82
t-Butanol	0.79	26	83
Group 3			
Cycloheptane	0.81	−12	118
n-Butanol	0.81	−90	118
s-Butanol	0.81	−89	100

The names of the substances in this table are not important now, and you do not need to remember them. They are good examples of substances that we cannot tell apart unless we measure all three properties: density, melting point, and boiling point.

In Group 1 of the table, the substances have the same boiling point and nearly the same melting point. It would be hard to measure these two properties carefully enough to see that they are different substances. However, a measurement of their densities would prove without question that they are different.

In Group 2 the substances have the same density and nearly the same boiling point, but can be told apart by their quite different melting points.

If you compared only their densities, you might conclude that the three substances in Group 3 are the same. If you also measured their melting points, you would probably decide that the second and third substances in this group are the same. If you compared their densities and boiling points but not their melting points, which would you conclude are the same? In fact, all three substances in Group 3 are different. That is why they were given different names when first discovered.

There are not very many examples of substances that are nearly the same in two of these three properties and yet differ in the third. It would be even harder to find two substances that have the same density, melting point, and boiling point but differ in some other property. If we can determine density, melting point, and boiling point, we can distinguish between almost all substances.

In many cases, the melting point and the boiling point of a sample of matter can be measured easily in the laboratory. However, some substances have boiling points so high that it is difficult to make them hot enough to boil. For example, table salt boils at 1,413°C. Others have boiling points so low that it is difficult even to make them cold enough to become liquid. The same experimental difficulties come up when we try to determine the melting points of some substances. For example, grain alcohol melts at –117°C.

Suppose you have a sample of a newly made substance. You wish to find out whether it is truly a new substance, different from all others, or a substance already known but made in a new way. If its boiling and melting points are too high or too low to measure easily, you must look for other characteristic properties that might help to distinguish it from similar substances.

29. **Which of the substances listed in Table 3.2 are solids, which are liquids, and which are gases at**
    a. **room temperature (20°C)?**   b. **50°C?**   c. **100°C?**

30. a. Would adding more ice to a Styrofoam cooler filled with canned drinks make the drinks colder?
    b. If not, how would adding more ice affect the drinks?
    c. Suggest an experiment to test your answer to part (b).

31. During a summer picnic the temperature rose from 25°C to 32°C.
    a. Did this change in temperature affect the temperature of canned drinks held in a Styrofoam ice chest?
    b. What change in the ice chest did this temperature change bring about?

32. The melting point of a few tiny crystals of BHT can be measured by placing the crystals in a small capillary tube held next to the bulb of a thermometer in a water bath. (See Figure D.) When the melting point is found by this method, the result agrees with the freezing point you found in Experiment 3.2, Freezing and Melting, using about 10 cm³ of the substance.
    a. Use Figure D to estimate the volume of BHT used in the experiment. The inner diameter of the capillary is about 0.5 mm.
    b. By what factor is the volume of BHT used in this experiment smaller than that used in Experiment 3.2?

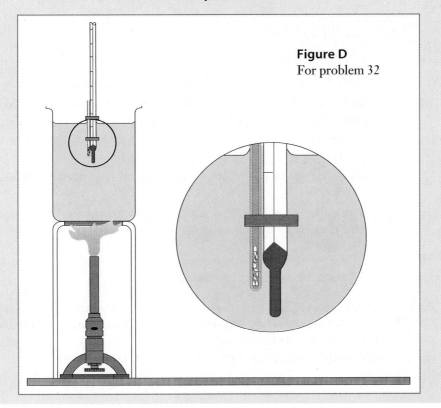

Figure D
For problem 32

33. Figure E shows a diagram of a double boiler. Why is a double boiler used to cook food that is easily scorched?

Figure E
For problem 33

Food to be cooked

Boiling water

Heater or flame

34. Object *A* has a mass of 500 g and a density of 5.0 g/cm³; object *B* has a mass of 650 g and a density of 6.5 g/cm³.
    a. Which object would displace the most liquid?
    b. Could object *A* and object *B* be made of the same substance?

35. A student uses water displacement to measure the volume of a small aluminum ball and then masses it on a balance. He finds that the sphere displaces 4.5 cm³ of water and that the mass of the sphere is 6.5 g.
    a. What value does the student obtain for the density of aluminum?
    b. How might you account for the difference between this value for the density of aluminum and the one given in Table 3.1?

36. How would you distinguish between unlabeled pint cartons of milk and of cream without breaking the seals?

37. A student has several different-size samples of substances *C* and *D*. She measures the masses and volumes of these samples and plots the graphs shown in Figure F on the next page. Which substance has the greater density? How do you know?

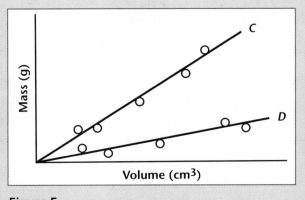

Figure F
For problem 37

38. How would you determine the density of ice? Could you determine the volume of the ice by melting it and measuring the volume of the resulting water?

39. In Table 3.1, why are the pressure and temperature stated for the densities of gases and not stated for the densities of solids and liquids?

40. A cylinder is closed with a tight-fitting piston 30 cm from the end wall (Figure G). The cylinder contains a gas with a density of 1.2×10^{-3} g/cm^3. The piston is pushed in until it is 10 cm from the end wall. If no gas escapes, what is the density of the compressed gas? What is your reasoning?

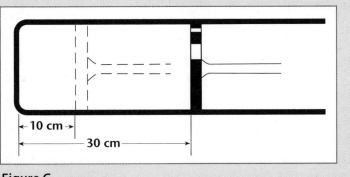

Figure G
For problem 40

41. Does the density of air change when it is heated
    a. in an open bottle?
    b. in a tightly stoppered bottle?

THEMES FOR SHORT ESSAYS

1. Ice floats on water. Suppose it did not. What would happen to life in lakes and rivers? Write a short science-fiction story on this subject.

2. Write a short mystery story in which a criminal tries to pass off a small statue of gold-plated lead as solid gold. Have the detective uncover the plot without damaging the statue.

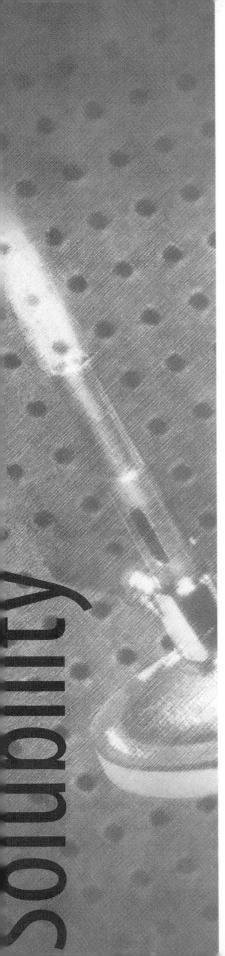

Chapter 4
Solubility

EXPERIMENT

4.1 Dissolving a Solid in Water

You know from daily experience that sand and chalk do not dissolve in water, but that sugar and table salt do. Of course, these are only qualitative observations—that is, observations that do not involve measurement. Are you sure that not even a tiny amount of chalk or sand dissolves in a gallon of water? Can you dissolve as much salt or sugar as you wish in a glass of water?

Many solutions are colorless—there is nothing to be seen once the solid is dissolved. To make things visible, we shall begin our quantitative study of solutions with a solution that has color.

Place 5.0 cm^3 of water in one test tube and 20.0 cm^3 of water in another. Add 0.30 g of orange solid (or 0.15 g of blue solid) to each tube, then stopper and shake thoroughly.

CAUTION: Be careful not to get any of the solution on your hands. If some spills, wash thoroughly with water.

- Did all of the solid dissolve in both test tubes?
- Do you think that each cubic centimeter of solution contains the same mass of dissolved material?
- Both tubes contain the same mass of dissolved material, but is the shade of the color the same in both tubes?
- Will another 0.30 g of the orange solid (or 0.15 g of the blue solid) dissolve in each of the test tubes? Try it, but be patient!

The solution may cool as the solid dissolves. If so, try to keep the temperature fairly constant by warming the test tube with your hand.

- Did 0.60 g of the orange solid (or 0.30 g of the blue solid) dissolve as well in 20.0 cm^3 as in 5.0 cm^3 of water?
- Is the color uniform in each solution?
- Is the shade of the color the same in both solutions?

Add another 0.30 g of the orange solid (or 0.15 g of the blue solid) to each solution.

- How much do you have in each test tube?
- What do you observe?
- How much orange (or blue) solid do you think there would have to be in the 20.0-cm^3 test tube before the solid stopped dissolving? Test your prediction.

A solution in which no more solid can be dissolved is called a *saturated solution*.

4.2 Concentration

In the last experiment, the color changed as you added more solid, but the color was uniform throughout each test tube. This uniformity of color suggests that the solid dissolved uniformly. That is, a cubic centimeter of water in a given solution contained the same mass of orange (or blue) material as any other cubic centimeter of water in that solution. For example, if you used the orange solid, at the beginning of the experiment one test tube contained

$$\frac{0.30 \text{ g}}{5.0 \text{ cm}^3} = 0.060 \text{ g/cm}^3$$

of material and the other test tube contained

$$\frac{0.30 \text{ g}}{20.0 \text{ cm}^3} = 0.015 \text{ g/cm}^3.$$

The mass of solid dissolved per unit volume of liquid is called the *concentration* of the solution. The unit of concentration is the same as that of density, g/cm^3. However, in the case of density, the mass and the volume refer to the same substance. In the case of concentration, the mass refers to the dissolved solid (called the *solute*) and the volume refers to the liquid (called the *solvent*).

To avoid confusion and for convenience, concentrations are often given in $g/100 \text{ cm}^3$. For example, a concentration of 0.015 g/cm^3 means that 0.015 g of solute is dissolved in 1 cm^3 of water. Hence, a volume of 100 cm^3 of water will contain

$$100 \times 0.015 \text{ g} = 1.5 \text{ g}$$

of solute. Thus, the concentration of the solution given in $g/100 \text{ cm}^3$ is $1.5 \text{ g}/100 \text{ cm}^3$. In general, to express the concentration of a solution in $g/100 \text{ cm}^3$, multiply the concentration in g/cm^3 by 100.

1. For Experiment 4.1, Dissolving a Solid in Water, calculate the concentration of the solutions in g/cm^3 and in $g/100 \text{ cm}^3$ after the addition of each sample of solid.

2. What was the largest concentration you were able to make in Experiment 4.1, Dissolving a Solid in Water?

3. A mass of 25.0 g of sugar is dissolved in 150 cm^3 of water. What is the concentration in
   a. g/cm^3?
   b. $g/10 \text{ cm}^3$?
   c. $g/100 \text{ cm}^3$?

EXPERIMENT
4.3 Comparing the Concentrations of Saturated Solutions

From the results of dissolving the orange or blue solid (Experiment 4.1, Dissolving a Solid in Water), you know that there is a point at which no more solute will dissolve in the solvent. The solution then has the largest possible concentration and is called a saturated solution, as stated in Section 4.1.

To find the concentration of a saturated solution, you could add the solid a tiny amount at a time and see whether it dissolves. A better method is to add a large mass of solid to a solute and shake the container until you judge that no more solid will dissolve. Then you can pour off some of the clear liquid and find the concentration.

Try dissolving 5 g of two solids in separate test tubes, each containing 5 cm³ of water. Stopper the test tubes, and shake them vigorously for several minutes until the solution is saturated. If the tube cools during the process, keep it warm with your hand.

• Does one sample of solid appear to be more soluble in water than the other?

To find the concentrations of the two saturated solutions, you can evaporate the liquid from a known mass of each solution. Subtracting the mass of the remaining dry solid from the mass of the solution will give you the mass and, therefore, the volume of the water of your sample. This will give you the data you need to calculate the concentration of the saturated solution. You can do the experiment for one solution while some of your classmates work with the other solution.

Pour almost all the saturated solution into a previously massed evaporating dish. Be careful not to pour out so much solution that undissolved solid is carried over from the test tube into the dish.

After finding the total mass of dish and solution, slowly evaporate the saturated solution to dryness over a flame, as shown in Figure 4.1, and find the mass of the remaining solid.

CAUTION: Always wear safety glasses when you use a burner.

Figure 4.1
Evaporating a solution in an evaporating dish heated over a microburner. If the liquid spatters, it should be heated more slowly by moving the burner to one side so that the flame heats only one edge of the dish.

Be careful to heat the solution very slowly so that solid does not spatter out of the dish. Keep watching the dish, and move the flame away whenever spattering begins.

- What is the mass of the solid and the mass of the water in which it dissolved?
- What is the volume of the water?
- What is the concentration of each of the saturated solutions?
- How do your results compare with those of your classmates?

The concentration of a saturated solution is called the *solubility*. In this experiment you found the solubilities of two substances in the same solvent, namely water.

Solubility is independent of the mass of the sample from which it is found. It is a characteristic property of the combination of the solute and the solvent.

4. Mario wishes to construct a table that lists the solubility in water of several substances. From various sources he finds the following data for solubilities at 0°C.
   a. Boric acid   0.20 g in 10 cm³ of water
   b. Bromine   25 g in 600 cm³ of water
   c. Washing soda   220 g in 1,000 cm³ of water
   d. Baking soda   24 g in 350 cm³ of water
   Find the solubility of each substance in g/100 cm³ of water.

5. From your answers to problem 4, find the largest mass of each substance that will dissolve in 60 cm³ of water.

6. Suppose that 200 cm³ of a saturated solution of potassium nitrate were left standing in an open beaker on your laboratory desk for three weeks. During this time most of the water evaporated.
   a. Would the mass of potassium nitrate dissolved in the solution change?
   b. Would the concentration of the potassium nitrate solution change during the three weeks?

EXPERIMENT
4.4 The Effect of Temperature on Solubility

In the last experiment, you tried to keep the temperature of the solution constant (by warming the test tube with your hand if it cooled). How will the solubility of different substances be affected by the temperature of the liquid? Remember that solubility is the maximum mass of a solid that will dissolve in a given volume of liquid.

To find out, add 10 g of two solids to two test tubes, each containing 10 cm³ of water. Place both test tubes in a large beaker of water, and stir the solutions for several minutes until they are saturated. Now heat the beaker, stirring both solutions constantly, until the water in the beaker is near boiling.

CAUTION: Always wear safety glasses when you use a burner.

- What do you observe?
- Do the solubilities of the substances appear to change equally or differently as the temperature of the water is increased?
- What do you predict will happen if you remove the burner and cool both test tubes together in a beaker of cold water? Try it.

Figure 4.2 shows the result of an experiment with potassium sulfate. The solubility at different temperatures was measured by the same method you used in Experiment 4.3, Comparing the Concentrations of Saturated Solutions. The solubility is expressed as the mass in grams of the substance that is dissolved in 100 cm³ of water to give a saturated solution or, to put it another way, the maximum mass of the substance that can be dissolved in 100 cm³ of water.

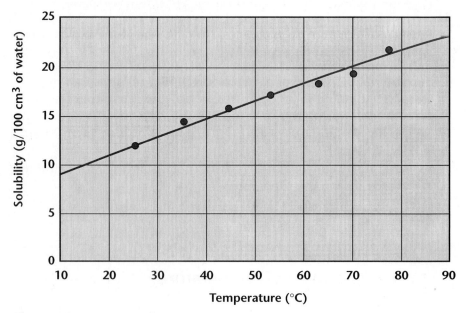

Figure 4.2
A graph of the solubility of potassium sulfate as a function of temperature. Note that the line does not extend through all of the points (Section 3.3).

Suppose we dissolve 20 g of potassium sulfate in 100 cm³ of water at 80°C, and then cool the solution to room temperature, 25°C. Figure 4.2 shows that the water can hold only 12 g of potassium sulfate in solution at this temperature. Therefore, during the cooling process, small crystals of solid potassium sulfate begin to appear in the solution at about 70°C. As cooling continues, more crystals are produced in the solution, and they sink to the bottom. A solid that crystallizes out of a saturated solution in this manner is called a *precipitate*. In this case, the mass of potassium sulfate that will precipitate out of solution and collect at the bottom will be 20 g – 12 g = 8 g.

Figure 4.3 shows solubility as a function of temperature for several common substances, all plotted together on the same graph. These

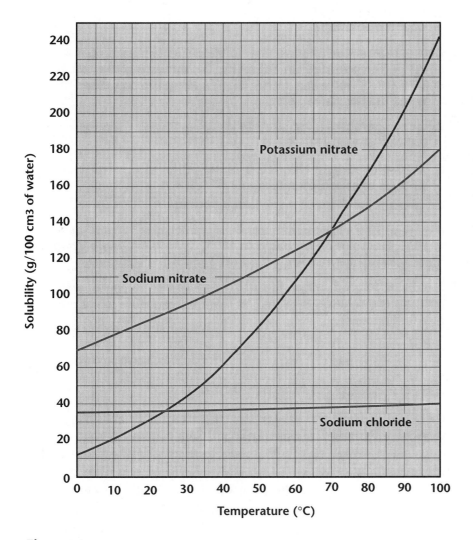

Figure 4.3
Solubility curves of different substances in water as a function of temperature.

curves clearly show that the way the solubility of a substance changes with temperature is a characteristic property that can help to distinguish between different substances. You can see from the graph, for example, that the solubilities of potassium nitrate and sodium chloride (ordinary table salt) are very nearly the same at room temperature (about 25°C) but are widely different at high and low temperatures.

7. If you plotted the data from Figure 4.2 on Figure 4.3, where would the curve be found?

8. Suppose you have a saturated solution of potassium sulfate at room temperature. From Figure 4.2, what do you predict will happen if you
   a. heat the solution?
   b. cool the solution?

9. Give a reason for heating the solutions in Experiment 4.4, The Effect of Temperature on Solubility, by immersing the test tubes in a beaker of hot water rather than heating the test tubes directly over a burner flame.

†10. What temperature is required to dissolve 110 g of sodium nitrate in 100 cm³ of water?

11. a. If 20 g of sodium chloride is dissolved in 100 cm³ of water at 20°C, is the solution saturated?
    b. How do you know if a solution is saturated?

†12. A mass of 30 g of potassium nitrate is dissolved in 100 cm³ of water at 20°C. The solution is heated to 100°C. How many more grams of potassium nitrate must be added to saturate the solution?

13. A mass of 10 g of sodium nitrate is dissolved in 10 cm³ of water at 80°C. As the solution is cooled, at what temperature will a precipitate first appear?

4.5 Wood Alcohol and Grain Alcohol

Most rocks and metals, and many other materials, are so slightly soluble in water that we cannot measure the very small amounts that dissolve. Water, however, is not the only liquid. Perhaps some substances that hardly dissolve at all in water will dissolve easily in other liquids. If so, we can use the different solubilities of substances in these liquids to distinguish between them, as we did with materials soluble in water. We shall first investigate wood alcohol and grain alcohol, two common liquids. Then we shall see if there are other solvents that will further increase our stock of tools for investigating matter.

Wood alcohol, as its name implies, was first made from wood. The ancient Syrians heated wood in order to obtain the liquids and tars that resulted. The watery liquids (including the alcohol) were used as solvents and as fuel for lamps. The tars were used to fill the seams in boats, to preserve wood against rot, and as mortar for bricks.

The method used by the Syrians in making these substances was quite crude. Lengths of wood were stacked closely in a dishlike depression in the top of a mound of earth. A drain ran from the middle of the depression to a collection pit. After the wood was covered with green branches and wet leaves, a fire was started inside the pile. As this fire smoldered, watery liquids and tars drained off from the pile and collected in the pit. Later it was discovered that the watery liquids could be separated. One of these liquids was wood alcohol.

Grain alcohol can be made by fermenting grains, such as corn, barley, and rye, as well as by fermenting grapes and other fruits. Fermentation is the process that goes on naturally when fruit juice or damp grain is stored with little exposure to air. Gas bubbles out of the liquid, and what remains boils at a temperature lower than the boiling point of water. As had been discovered long before the beginning of recorded history, this liquid contains a new substance, different from water, which was used as a beverage and as a medicine.

By the twelfth century, the wine from fermented grapes was boiled and the vapor collected and allowed to cool. The condensed liquid was described as the "water that burns." It was named *alcohol vini*, or "essence of wine," and later came to be called grain alcohol.

If you measure the densities of the alcohols from different grains and fruits, you find no difference between them. These alcohols also have the same boiling point and melting (or freezing) point. In fact, they are all the same substance. Similarly, the alcohols (or essences) produced from different kinds of wood are all the same: all have the same density, boiling point, and melting point. Since the liquids obtained from wood and fermented grain are both called alcohol, you might think that they are the same substance. But from an examination of Table 4.1, you see that they are indeed different. Though their densities are nearly the same, their melting points and boiling points differ enough so that there can be no possibility that they are the same substance. Today, wood alcohol is called *methanol*, and grain alcohol is called *ethanol*.

Table 4.1 Some Characteristic Properties of the Most Common Alcohols

	Density (g/cm³)	Melting point (°C)	Boiling point (°C)
Wood alcohol (methanol)	0.79	–94	65
Grain alcohol (ethanol)	0.79	–117	79

Both alcohols can be used as fuels. The fuel in a laboratory alcohol burner should contain mostly ethanol. Ethanol is the important ingredient in alcoholic beverages. Methanol, on the other hand, is poisonous. Some methanol, or other poison, is usually added to burner fuel to make it undrinkable. Ethanol treated this way is said to be denatured. Both alcohols will dissolve many substances that are insoluble in water, and both have been used for centuries as solvents.

During the nineteenth century, substances with properties similar to those of methanol and ethanol were made in the laboratory. They were also called alcohols. A common alcohol is *isopropanol*, also known as rubbing alcohol. Isopropanol also dissolves many substances insoluble in water. You will use isopropanol in the next experiment.

EXPERIMENT
4.6 Isopropanol as a Solvent

Sugar and citric acid look the same. They are both white. Sugar has a density of 1.58 g/cm^3, and citric acid has a density of 1.67 g/cm^3. These densities differ by only 5 percent. You would find it difficult to determine the volume of an irregular piece of either material to within 5 percent. Even if you made such volume measurements, the calculated values for the densities would be reliable only to 5 percent. Thus, from a density determination, it will be difficult to distinguish between these two substances. Their solubility in water is also not much help because both substances are very soluble in water.

Suggest a way to distinguish between these two substances by using their solubility in isopropanol. Have your teacher approve your plan before you begin.

CAUTION: Do not inhale isopropanol vapor.

- How do the solubilities of sugar and citric acid in isopropanol help to distinguish between these substances?

You can also test the solubilities of other substances, such as TOP and baking soda.

- Does TOP dissolve in water? In isopropanol?
- Does baking soda dissolve in water? In isopropanol?

You can see how valuable solubility in different solvents is for distinguishing between substances. Suppose you have samples of two solids that look alike and dissolve about equally well in one solvent but very differently in a second solvent. You can be certain that the two solids are different substances. You saw an example of this in the case of sugar and citric acid in isopropanol.

Likewise, two similar liquids can be distinguished if a substance dissolves to a much greater extent in one liquid than in the other. You saw examples of this in the case of the solubility of TOP or sugar in water and isopropanol.

Solubility is another characteristic property of a substance, a tool that you can add to your list of ways to distinguish between substances. With the addition of solubility, you do not always have to measure the density, melting point, or boiling point to distinguish between substances. The more tools you have for distinguishing between substances, the easier your job becomes.

14. Two solids look alike and are both insoluble in rubbing alcohol. A student, whose results are reliable to 5 percent, reports the solubilities of the two solids in water, as in the following table. Are these solids the same substance? Explain your answer.

	Solubility (g/100 cm³)	
Solid	0°C	100°C
A	73	180
B	76	230

15. a. Which of the following substances, X, Y, and Z, do you think are the same?
    b. How might you test them further to make sure?

Substance	Density (g/cm³)	Melting point (°C)	Boiling point (°C)	Solubility in water at 20°C (g/100 cm³)	Solubility in methanol at 20°C
X	1.63	80	327	20	insoluble
Y	1.63	81	326	19	insoluble
Z	1.62	60	310	156	insoluble

4.7 Sulfuric Acid

Another solvent useful in distinguishing between different substances was first produced more than a thousand years ago by heating a soft rock called green vitriol. The vapors produced by heating the rock were

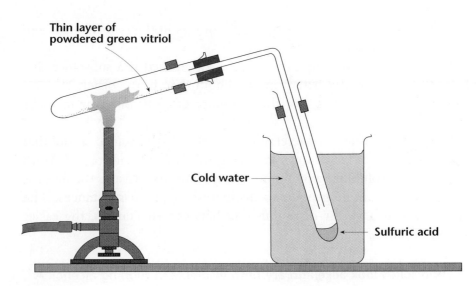

Thin layer of
powdered green vitriol

Cold water

Sulfuric acid

Figure 4.4

A classroom adaptation of the ancient method of producing oil of vitriol.

cooled and condensed to form an oily liquid that was called oil of vitriol (see Figure 4.4). Its modern name is *sulfuric acid*.

The ancient method of producing sulfuric acid has been abandoned in favor of a far more effective process that begins with sulfur. As a result, the acid is readily available in large quantities. The yearly production of sulfuric acid in the United States is about 4×10^{10} kg. Sulfuric acid is of great importance in industry and is used in the manufacture of many other substances.

One of the useful properties of sulfuric acid is its ability to dissolve some substances that will not dissolve in water. In many cases, when a substance is dissolved in sulfuric acid, a gas is given off. We shall investigate two such gases in the next experiment.

EXPERIMENT
4.8 Two Gases

One of the substances that produce a gas when dissolved in sulfuric acid is magnesium. To collect the gas, you can use the same apparatus as in Experiment 3.11, The Density of a Gas.

CAUTION: Sulfuric acid is highly corrosive, so try not to spill it or get any on your hands, clothing, or books. If you do, wash it off immediately with water and tell your teacher. Always wear safety glasses when you work with acids and gases.

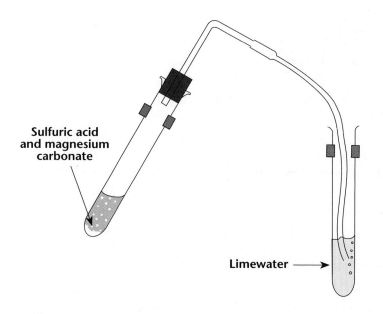

Figure 4.5
The limewater test. Gas from the test tube on the left is bubbled through limewater in the test tube on the right.

Use about half a test tube of acid and five 7-cm lengths of magnesium ribbon to produce several test tubes of gas.

- Should you discard the gas collected in the first test tube? Why?

Try lighting tubes of gas, using first a burning splint and then a glowing one. Try it holding some of the tubes upside down and some right-side up.

- Does the gas burn?
- Is the gas more dense or less dense than air?

Try bubbling a little of the gas into limewater (Figure 4.5).
Repeat the experiment, using magnesium carbonate instead of magnesium.

4.9 Hydrogen

The gas you made by dissolving magnesium in sulfuric acid was first produced many centuries ago by the action of sulfuric acid on metals. It was called "flammable air" because it burned. Later it was discovered that when flammable air burns, it produces another gas, which can easily be condensed to a liquid. This liquid was found to boil at 100°C, freeze at 0°C, have a density of 1.00 g/cm^3, and dissolve the same substances that water does. In fact, all the properties of this liquid were the same as those of water, and so it was considered highly unlikely that it could be anything else. Because it produced water when it burned, the gas was named *hydrogen* about 200 years ago. "Hydro" is a prefix that means "water." "Gen" is a suffix that means "producing" or "generating."

However, other gases also produce water when burned and, like hydrogen, are less dense than air. For many years, these gases were thought to be the same as hydrogen. Marsh gas, now called *methane*, is an example that has been known for thousands of years. It is often produced by vegetable matter decaying at the bottoms of lakes or marshes, where it slowly bubbles up to the surface of the water. But this gas does not have nearly as low a density as hydrogen, and so it must be a different substance. It was not until the density of gases could be measured with reasonable accuracy that marsh gas was found to be a different substance from hydrogen.

16. **A student might try to find the density of hydrogen by the method you used in Experiment 3.11, The Density of a Gas. There you collected about 400 cm³ of gas.**
    a. **What would be the mass of this volume of hydrogen?**
    b. **What difference in mass would there be between the total mass of test tube, acid, and magnesium at the beginning and the total mass of test tube and contents at the end of the action?**
    c. **Could you accurately measure this difference in mass on your balance?**

4.10 Carbon Dioxide

The gas that you produced by dissolving magnesium carbonate in sulfuric acid is called *carbon dioxide*. It is a part of the air we exhale at every breath and is also produced by a variety of substances when they are placed in sulfuric acid. The density of carbon dioxide is 1.8×10^{-3} g/cm³, which is much greater than that of air. Since carbon dioxide does not burn, it makes a good fire extinguisher. It simply smothers the fire like a blanket.

Solid carbon dioxide, called *dry ice*, evaporates directly into carbon dioxide gas without first turning into a liquid. But, like melting ice, it stays cold until all the solid is gone. Ice melts at 0°C, but dry ice evaporates at a much lower temperature (at −78.5°C). It remains at this temperature as it evaporates, and for this reason it is often used to cool things down to very low temperatures.

EXPERIMENT
4.11 The Solubility of Carbon Dioxide

If the gases you have investigated so far had been very soluble in water, you could not have collected them over water. Much of the gases would have dissolved as they were produced. Nevertheless, no gases are completely insoluble in water. In this experiment you will measure the solubility of carbon dioxide by dissolving a known mass of the gas in a known volume of water.

Figure 4.6 shows the equipment you will use to generate carbon dioxide gas and to collect it by water displacement.

CAUTION: Always wear safety glasses when you work with gases.

Before you begin to generate the gas, you need to prepare the known volume of water in which you are going to dissolve it. An easy way of doing this is to fill completely the collecting bottle with cold tap water and pour it into the container in which you will later invert the collecting bottle. This container is shown on the right in Figure 4.6. To find

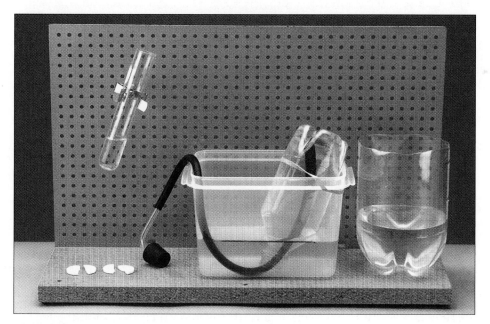

Figure 4.6
Equipment used to collect a full bottle of carbon dioxide and to measure its solubility. The large test tube contains 10 cm³ of water and the inverted collecting bottle is filled with water. The volume of water in the container on the right is the same as the volume of the collecting bottle.

the volume of the collecting bottle, you can use your 250-mL beaker that has graduation marks.

• What is the volume of the collecting bottle?

Two Alka-Seltzer tablets (four halves) placed in the test tube will generate enough gas to fill the collecting bottle. After adding the tablets, quickly put the rubber stopper in place and carefully insert the delivery tube into the top of the collecting bottle. While you do this, the carbon dioxide will force the air out of the test tube and the rubber tube so it will not be collected.

• How will you know when you have collected a full bottle of carbon dioxide?

When the collecting bottle is completely filled with gas, invert it in the container having the same volume of water (Figure 4.7). Be sure not to lose any gas in the process.

Gases, like most substances, do not dissolve rapidly. Try moving the inverted plastic bottle back and forth in the larger container in order to cause a little water to enter its mouth. Be careful to keep the mouth of the bottle against the bottom of the container so that no bubbles of carbon dioxide escape. As more gas dissolves and water rises into the

Figure 4.7
A bottle full of gas ready to be placed in the container of water. To prevent the gas from escaping, be sure to hold your hand over the mouth of the bottle until it is under water. (Blue color was added to the water for contrast.)

Figure 4.8
A bottle full of gas placed in the container of water. Shake the inverted bottle back and forth vigorously so that the gas and water mix well. Hold the mouth of the bottle firmly against the bottom of the plastic container to prevent any gas from escaping. (Blue color was added to the water for contrast.)

collecting bottle, you can shake it back and forth more and more vigorously. The shaking also mixes the water inside and outside the collecting bottle, ensuring that the gas is dissolved in all the water.

- After 10 to 15 minutes of shaking, does the carbon dioxide continue to dissolve?
- How will you know when the water has become saturated with carbon dioxide?

In order to measure the volume of gas that dissolved, the bottle must be turned upright without losing any saturated solution from inside the bottle. This can be done by quickly inverting the entire assembly shown in Figure 4.8 over a large sink. Only the solution from the container that is outside the bottle will spill into the sink.

- When the water becomes saturated, what is the volume of carbon dioxide gas that does not dissolve?
- What is the volume of carbon dioxide that does dissolve?
- Use the class data of Experiment 3.11, The Density of a Gas, to find the mass of carbon dioxide that dissolved in the saturated solution.
- What is the volume of water in the saturated solution?
- What is the solubility of carbon dioxide in g/100 cm^3 of water?

---

17. **Experiment 3.11, The Density of a Gas, was done twice, but with one tablet and 15 cm^3 of water in each case. When the rubber tube was placed as shown in Figure 3.10, 435 cm^3 of gas was collected. When the tube reached only slightly beyond the mouth of the bottle, only 370 cm^3 of gas was collected. The change in mass of the reactants was the same in both cases.**
    a. **What volume of gas dissolved in the water?**
    b. **Use class data of Experiment 3.11 to find what mass of gas dissolved in the water.**

4.12 The Solubility of Gases

Table 4.2 shows the solubilities of several gases in water. It is clear from the table that the solubility of gases varies greatly. Gases of low solubility can be collected by water displacement. Others, such as hydrogen chloride, cannot be collected over water. Hydrogen chloride was prepared over two hundred years ago by the French chemist Antoine Lavoisier by reacting sulfuric acid with table salt. He collected the gas by displacement of mercury.

You can see in Table 4.2 that the solubility of nitrogen is small compared to those for hydrogen chloride and ammonia, which are two of the most soluble gases known. The solubility of all the gases in Table 4.2 decreases as the temperature increases. This behavior is different from that of solids, which you observed in Experiment 4.4, The Effect of Temperature on Solubility, and in Figures 4.2 and 4.3. The solubility of most solids increases with temperature but that of most gases diminishes.

Table 4.2 The Solubility of Gases in Water at Various Temperatures (in g/100 cm^3)

Temperature (°C)	Nitrogen	Oxygen	Carbon dioxide	Sulfur dioxide	Hydrogen chloride	Ammonia
0	2.4×10^{-3}	7.0×10^{-3}	0.34	23	82	90
20	1.9×10^{-3}	4.4×10^{-3}	0.17	11	72	53
40	1.4×10^{-3}	3.3×10^{-3}	0.10	5.5	63	32
60	1.1×10^{-3}	2.8×10^{-3}	0.07	3.3	56	14

[†]18. Fish prefer to live in cool water that has splashed over rocks rather than in warm water that is stagnant. Based upon information in Table 4.2, present an explanation for this observation.

19. Carbonated beverages are normally saturated solutions of carbon dioxide that have other substances dissolved in them to provide color and flavor. When a carbonated beverage is left standing in an open glass, it becomes "flat." Give an explanation for this observation.

20. Carbonated soft drinks can be transported in the sun and stored at room temperature. When they are opened cool, they fizz, showing that the solution is saturated in carbon dioxide.

   a. What do you think happens to the carbon dioxide in the water when the bottles are in the sun?

   b. Why do glass bottles containing carbonated drinks sometimes explode when left in the sun for too long?

   c. What happens to the carbon dioxide when the bottles cool down again?

4.13 Acid Rain

When some gases, such as carbon dioxide, sulfur dioxide, and hydrogen chloride, are dissolved in water, the properties of the solution are very different from those of water alone. These solutions have a sour taste and are able to dissolve minerals and metals to a much greater extent than water. They are called *acidic solutions.*

Raindrops dissolve carbon dioxide as they pass through the atmosphere, forming an acidic solution known as *carbonic acid.* Limestone and chalk are much more soluble in carbonic acid than in water. When rain water seeps through the soil and reaches a layer of limestone, it slowly dissolves the limestone, leaving an empty space. When the soil above this empty space can no longer support itself, it collapses, as shown in Figure 4.9.

Carbon dioxide is produced and added to the atmosphere when coal, oil, wood, and gasoline are burned. Consumption of these fuels is increasing worldwide. Thus, the concentration of carbon dioxide in the atmosphere is increasing.

Much of the coal that we burn as fuel contains sulfur. When sulfur is burned, sulfur dioxide is produced. As you can see in Table 4.2, sulfur dioxide has a higher solubility in water than does carbon dioxide. Water

Figure 4.9
A "sinkhole" in central Florida. Limestone below the top layer was dissolved in water containing much carbon dioxide. (*AP/Wide World Photos*)

Figure 4.10
A statue of Civil-War General William F. Draper in Milford, Massachusetts. The statue, made in 1912 and photographed in 1987, has been damaged by acid rain. (*Courtesy, Center for Conservation and Technical Studies, Harvard University Art Museums*)

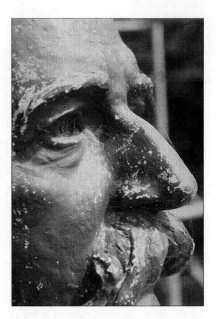

falling to the earth that contains dissolved sulfur dioxide is called *acid rain*.

Acid rain dissolves the surface of stone used in buildings and statues, as shown in Figure 4.10. It can also damage or kill plants, as Figure 4.11 shows. If enough acid rain or snow accumulates in lakes and ponds, it is harmful to the animals living there.

4.14 Drinking Water

Most people get their drinking water from wells, rivers, or lakes. This water was once rain or snow that travelled great distances through the air, over rocks and land, and through soil, dissolving many substances along the way. In addition to dissolved minerals, drinking water may contain fertilizers, herbicides, pesticides, and other substances made by people.

The concentrations of substances dissolved in drinking water are very low. If they were not, the water would be harmful and have a strange taste. As an example, the concentration of ordinary table salt in your water is probably less than 0.005 g/100 cm^3. The concentrations of several substances in the drinking water of three cities are listed in Table 4.3.

The concentrations listed in Table 4.3 are very low compared to the solubilities that you measured and observed in earlier

Figure 4.11
Spruce trees on Whiteface Mountain in New York that have been killed by acid rain. (*USDA Forest Service*)

sections. Also, the concentrations listed vary from substance to substance and city to city.

Sodium chloride, iron chloride, and calcium carbonate are examples of substances that are not very toxic but affect the taste of water. Water that contains too much iron has a strange taste and leaves a rust-colored stain in containers. Water containing excessive amounts of calcium carbonate is said to be *hard water*. This water has a mineral-like taste and soaps do not lather or clean effectively in it.

Table 4.3 Analysis of City Water Supplies

Substance	Concentration (g/100 cm³)		
	Belmont, Mass.	Denver, Co.	Memphis, Tenn.
Arsenic chloride	$< 2.4 \times 10^{-6}$	$< 2.4 \times 10^{-6}$	$< 1.2 \times 10^{-6}$
Calcium carbonate	1.2×10^{-3}	2.2×10^{-3}	1.2×10^{-3}
Iron chloride	1.4×10^{-6}	1.4×10^{-5}	9.6×10^{-5}
Lead nitrate	$< 2.8 \times 10^{-7}$	$< 2.8 \times 10^{-7}$	1.4×10^{-7}
Mercury chloride	$< 2.7 \times 10^{-8}$	$< 1.4 \times 10^{-8}$	$< 2.7 \times 10^{-8}$
Sodium chloride	2.0×10^{-3}	5.1×10^{-3}	2.5×10^{-3}
Potassium nitrate	2.6×10^{-4}	1.8×10^{-4}	1.8×10^{-4}

Other substances are important because small concentrations of them in drinking water can be dangerous to people's health. Substances that contain arsenic, lead, and mercury are toxic substances that are harmful if present in too high a concentration in drinking water.

In Experiment 4.8, Two Gases, you saw how sulfuric acid dissolved magnesium metal. If too much acid rain becomes a part of the water supply, drinking water becomes able to dissolve metals. When acidic drinking water passes through lead pipes or pipes having joints sealed with lead solder, some of this lead dissolves in the drinking water.

Drinking water with a high enough concentration of lead can cause damage to the brain and other parts of the nervous system. Lead seems to affect young children more than adults. Perhaps this is because lead is not eliminated from the body quickly, as many other toxic substances are. Instead, it accumulates in the body, so that its concentration increases as time passes.

21. Which city listed in Table 4.3 has the hardest water?

22. A typical person drinks about two liters (2,000 cm³) of water a day. How much sodium chloride would a person in Memphis take in from drinking water each day? How much in a year?

23. Large deposits of sodium chloride or potassium nitrate near the surface of the earth are found mostly in deserts. Why?

FOR REVIEW, APPLICATIONS, AND EXTENSIONS

24. Each of four test tubes contains 10 cm^3 of water at 25°C. The following masses of an unknown solid are placed in the test tubes: 4 g in the first, 8 g in the second, 12 g in the third, and 16 g in the fourth. After the tubes are shaken, all of the solid has dissolved in the first two tubes, but some undissolved solid remains in the other two tubes.

   a. What is the concentration of the solid in each of the first two tubes?

   b. What can you say about the concentration of the solid in the second two tubes?

25. a. Which of the substances shown in Figure 4.3 could be the unknown solid of problem 24?

   b. If the unknown is indeed the substance you named in answer to part (a), what will happen if the solution in each test tube is cooled to 10°C?

26. The solubility of a substance in water at various temperatures is shown below. What would you expect the solubility to be at 60°C? At 100°C? Explain.

Temperature (°C)	Solubility (g/100 cm^3)
25	5
50	10
75	15

27. In many localities, after a kettle has been used for some time for boiling water, a flaky solid appears on the inside-bottom and on the sides of the kettle that have been in contact with the water. How do you account for the presence of this "boiler scale"?

28. There are two kinds of felt-tip (magic-marker) pens. Some are labeled "permanent" and some are labeled "water color."

   a. What does the label tell you about the solubility in water of the dye in the two inks?

   b. Do you think the liquid in both inks is water?

29. In dry cleaning, a garment is sprayed with liquids that dissolve various stains. Often a brightly colored cotton shirt carries a label: "Dry clean only—colors may bleed when washed."

    a. What does such a label tell you about the solubility of the dye in hot water containing a detergent or soap?

    b. Do you think the dye dissolves in cold tap water?

30. When baking soda is placed in sulfuric acid, a gas is produced whose properties you studied in Experiment 4.8, Two Gases. When you place washing soda in hydrochloric acid, you also produce a gas. How could you determine whether this gas is the same gas you generated by dissolving baking soda in sulfuric acid?

31. Three samples of gas are tested for characteristic properties. Sample *A* does not turn limewater milky, is less dense than air, and burns. Sample *B* turns limewater milky, is more dense than air, and burns. Sample *C* turns limewater milky, is less dense than air, and burns. What can you conclude about these samples of gas?

32. An Alka-Seltzer tablet is dissolved in 10 cm^3 of water, and the gas is collected as in Experiment 3.11, The Density of a Gas. The volume of gas collected is 450 cm^3. When 50 cm^3 of water is used, the volume of gas collected is 400 cm^3. The tube was all the way up in the bottle in both cases.

    a. Why do you think less gas is collected when more water is used to dissolve the tablet?

    b. Would this make a difference in the density calculation?

33. If you have a certain amount of a solid to dissolve in water, you usually can hasten the process in various ways. Why do you think each of the following steps is effective in making the solid dissolve faster?

    a. Stirring the water

    b. Crushing the solid into smaller particles

    c. Heating the water

34. Water analyses usually give the concentrations of dissolved substances in parts per million (ppm). If the concentration of sodium chloride in drinking water was listed as 14 ppm, what would its concentration be in g/100 cm^3?

35. The concentration of calcium carbonate in Denver's drinking water is 2.2×10^{-3} g/100 cm^3 (Table 4.3). Yet, at 25°C the solubility of calcium carbonate in pure water is only 1.4×10^{-3} g/100 cm^3. How can the concentration be higher than the solubility?

THEME FOR A SHORT ESSAY

Write an episode in a mystery centered on the possibility of selectively dissolving some substances mixed in with others. For example, a secret message written in ink may be covered by a painting. Describe real (not fictitious) substances.

Chapter 5

The Separation of Mixtures

In this chapter we shall use the characteristic properties studied so far to work out a variety of methods for separating mixtures. These mixtures may be gases, liquids, or solids. We shall begin with a careful investigation of what happens when we distill a mixture of liquids.

EXPERIMENT
5.1 Fractional Distillation

In this experiment you will determine some of the properties of a mixture of liquids. Then you will distill the mixture and examine the properties of the fractions to see if you succeeded in separating the liquids that made up the original mixture.

Part A

- Can you tell just by looking at it that the liquid is a mixture?
- Does the liquid have an odor?

Dip a small piece of paper in the liquid, and with a match try lighting the liquid on the paper.

CAUTION: Always wear safety glasses when you work with a flame. Put a bucket of water on your table in case the paper burns.

- Does the liquid burn?
- What is its density?
- Does sugar dissolve in the liquid?

Part B

Use the apparatus shown in Figure 5.1 to distill 5 cm^3 of the mixture. Use a single collecting tube and heat the liquid just enough to keep it boiling. Record the temperature of the vapor from the boiling liquid every half-minute while it distills. Continue to boil the liquid almost to dryness.

Make a graph of the temperature of the vapor as a function of time.

- What do you conclude from your graph about the number of fractions you should collect to separate the different substances in the mixture?
- At what temperatures should you shift from one collecting tube to another?

Indicate on your graph the temperatures at which you decided to change collecting tubes.

Figure 5.1

Apparatus for the fractional distillation of a liquid. The thermometer bulb is close to the top of the test tube so that it measures the temperature of the vapor that condenses in the outlet tube. If there is more than one liquid in the boiling mixture, most of the high-boiling-point liquids will condense and flow back down the test-tube walls before they reach the upper part of the test tube. The thermometer in this apparatus is not used to measure the boiling temperature of the liquid, but serves to indicate when collecting tubes should be changed to receive different fractions.

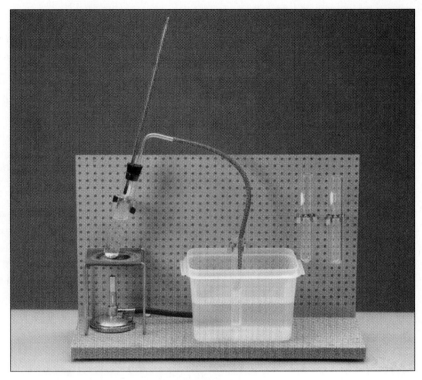

Part C

Now fractionally distill about 25 cm³ of the liquid. Label the test tubes containing the fractions, so that you can keep track of them throughout the rest of the experiment. Test each of the fractions for odor and flammability.

Part D

• Does sugar dissolve in fraction 1?

• What is the density of fraction 1?

Distill the fraction into a single test tube, recording the temperature of the vapor every half-minute until the fraction has nearly boiled away. Draw a vapor-temperature graph for fraction 1.

Part E

Repeat Part D for each of the other fractions.

Part F

Summarize your findings, and compare the odor, flammability, density, ability to dissolve sugar, and boiling point of each of the fractions and of the original mixture.

- What do you conclude about the composition of the fractions?
- Can you identify the substances in the mixture? (See Table 3.2, page 59.)
- What other tests might you make to help identify these substances?
- What do you think would happen if you were to re-distill each of the fractions separately?

†1. In what characteristic property must two liquids differ if a mixture of them is to be separated by fractional distillation?

2. The temperature-time graph shown in Figure A was made during the fractional distillation of a mixture of two liquids, E and F, and fractions were collected during the time intervals I, II, III, and IV. Liquid E has a higher boiling point than liquid F. What liquid or liquids were collected during each of the time intervals?

Figure A
For problem 2

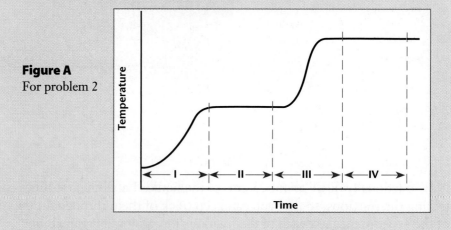

3. Juan boiled a liquid and recorded the temperature at 1-minute intervals until the liquid had nearly boiled away. How do you explain the shape of the curve he drew? (See Figure B.)

Figure B
For problem 3

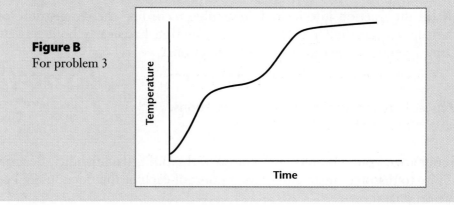

5.2 Petroleum

It is not always easy to separate a mixture of liquids into pure substances by fractional distillation. If the boiling points of the substances in a mixture are nearly the same, they will all boil off together. In a liquid mixture containing many substances, some are sure to have boiling points close together. When we distill such a mixture, each fraction is made up of a number of different substances whose boiling points are close together. The first part to condense in one fraction may contain some of the substances that condensed in the last part of the previous fraction. However, the fractions that are more widely separated in boiling range are less likely to contain the same substances. Petroleum is an example of such a mixture. The compositions of typical fractions distilled from petroleum are shown in Table 5.1.

Table 5.1 A Few of the Substances Found in Petroleum

| Substance | Density at 0°C (g/cm³) | Freezing point (°C) | Boiling point (°C) | Common products of petroleum | | |
				Fuel gas	Gasoline	Kerosene
Methane	7.16×10^{-4}	−182.5	−161	X		
Ethane	1.35×10^{-3}	−183	−88	X		
Propane	2.02×10^{-3}	−190	−43	X		
Butane	2.68×10^{-3}	−138	−0.5	X		
Pentane	0.626	−129	36			
Hexane	0.660	−94	69		X	
Heptane	0.684	−90	98		X	
Octane	0.703	−57	125		X	
Nonane	0.722	−51	151		X	X
Decane	0.730	−30	174		X	X
Undecane	0.741	−26	196			X
Dodecane	0.750	−10	216			X
Tridecane	0.755	−5.5	236			X
Tetradecane	0.765	5.5	254			X
Pentadecane	0.776	10	271			X
Hexadecane	0.773	18	287			X

There are many more substances in the above products and in the higher-boiling-point fractions not listed in the table. Among these substances are fuel oils, lubricating oils, waxes, asphalt, and coke (mostly carbon).

Petroleum is believed to be produced naturally from vegetable and animal matter at the bottoms of shallow seas and swamps. When tiny plants and animals die in the sea, they settle slowly to the bottom, where

Figure 5.2

A cross section of the earth's crust, showing how oil and natural gas are trapped in a porous rock layer by nonporous rock layers above and below. Note that well A produces only water, and well C produces only gas.

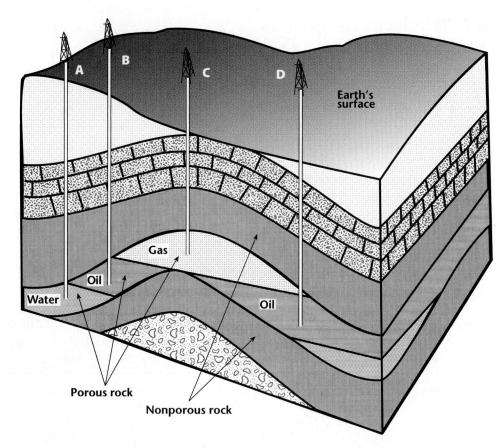

they become trapped in mud and sand. This sediment of mud, sand, and dead organisms slowly becomes thicker and thicker. In a million years, it may become hundreds of meters deep. Such layers of sediment are very heavy, and the lower layers are compressed so much that they turn into rock layers. During this time some of the body tissue of the entrapped organisms is changed into a viscous, sticky liquid that is a mixture of many thousands of different substances. This liquid is called *petroleum* or *crude oil*. It is slowly squeezed out of the sediment in which it forms and eventually spreads through porous rock layers like water in a sponge.

In the course of more millions of years, the ever-changing crust of the earth—buckling in some places, rising in others, and sinking in still others—slowly moves and compresses the rock layers that were on the ocean bottom. Sometimes the porous oil-bearing rock is covered by a layer of hard, nonporous rock that has been bent into a dome or arch, as shown in Figure 5.2. Then the oil will be trapped, and cannot by itself squeeze to the surface. If it were not trapped, much of it would wash away and be lost.

Most of the petroleum in the earth's crust is stored by nature under formations of nonporous rock, which trap the liquids below them. As

Figure 5.2 shows, natural gas (the low-boiling-point substances in petroleum) and salt water from the sea are often trapped along with the oil. The nonporous "cap rock" may be hundreds of meters thick. It is expensive and difficult to drill through all this rock to get to the petroleum below, and it is not easy to predict where oil is trapped. Deep and expensive wells often fail to reach oil or gas. Some wells produce nothing but salt water, whereas others remain dry.

Petroleum was first discovered where it seeped to the surface in shallow pools. Once exposed to the air, some of the lower-boiling-point substances slowly evaporated, leaving behind tarry, almost solid asphalt. These tars, as well as the liquid petroleum, were used in the ancient world as mortar and embalming material.

One of the ancient methods used to distill crude oil consisted of heating the oil in a copper urn with a wool "sponge" at the narrow mouth of the vessel. The vapors condensed in the sponge, which was squeezed out into containers from time to time. A variation of this method made use of a heavy wick of wool that led from the mouth of the urn into a collecting vessel. Such a wick was a crude form of condenser.

The widespread use of kerosene lamps over 100 years ago—and the more recent use of gasoline engines—created a new demand for petroleum. This led to improved methods of locating oil and drilling wells. Better equipment was also developed for fractionally distilling petroleum on a large scale (Figure 5.3).

Figure 5.3
A fractional distillation column at the Corpus Christi refinery. (*American Petroleum Institute*)

Figure 5.4

A simplified diagram of a fractionating column used in the fractional distillation of petroleum.

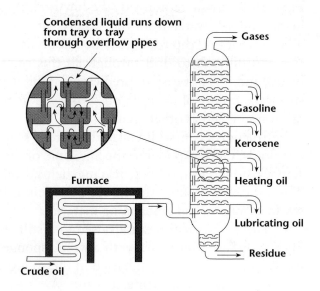

A simplified diagram of this equipment, called a *fractionating column*, is shown in Figure 5.4. Heated crude oil enters the column near the bottom. The column contains a series of horizontal trays. As vapors from the heated liquid pass up through the column, the high-boiling-point substances condense in the lower, hotter trays. As the vapors move upward, they bubble through the liquid in the trays. In each tray, the condensing vapors increase the concentration of the substances whose boiling points are higher than the temperature of the tray. The rising vapors therefore become richer in the low-boiling-point substances. Some of the liquid that condenses in each tray overflows into the tray below, where it becomes heated again and re-distills. Each tray thus boils a particular mixture at a particular temperature. The temperatures of the trays and the boiling points of the substances in them decrease as one goes higher up the column.

Different fractions leave the column at different heights (Figure 5.4). Fractions with which you may be familiar are gasoline, kerosene, diesel fuel, heating and lubricating oils, paraffin, and asphalt.

4. a. In Table 5.1, which fractions would be liquid at room temperature (20°C)? Which would be solids? Which would be gases?

   b. You can see from the table that pentane is not an ingredient in any of the common products listed. How can you account for this?

5. A sample of crude oil is boiled for several minutes. What change takes place in its density? (See Table 5.1.)

5.3 The Separation of Insoluble Solids

Solids with a density greater than that of water sink in water, and those with a lesser density float, provided they are insoluble. We can use this fact to separate a mixture of sawdust and sand. After stirring the mixture in water, we can skim off the floating sawdust. Then we can pour off the water and dry the sand.

This method is called *separation by flotation* and is widely used in industry to concentrate ores. For example, a common copper ore, copper sulfide, is usually found mixed with large amounts of worthless rock. The ore is finely ground, and then mixed with water and selected substances that produce a heavy foam when the mixture is agitated violently with air. The copper sulfide is contained in the floating foam; the ground rock settles. The foam is removed, and the copper sulfide is recovered.

EXPERIMENT
5.4 The Separation of a Mixture of Solids

Examine the mixture of solids supplied by your teacher. If one solid is soluble in water and the other is not, you can separate them easily by dissolving one and separating it from the other by filtering. You can do this in the following way: Put about 1.5 g of the mixture into a test tube, and add 5 cm³ of water. Stopper the test tube, and shake it for several minutes.

- Do you think either substance dissolved?

To find out, filter out the undissolved solid, as shown in Figure 5.5. Wash the precipitate on the filter paper by pouring an additional 10 cm³ of water into the funnel. You can now put about 5 cm³ of the clear liquid—the filtrate—into an evaporating dish and boil it to dryness.

- Have the two substances been separated?

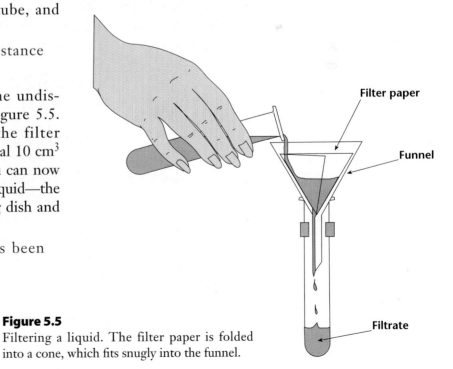

Figure 5.5
Filtering a liquid. The filter paper is folded into a cone, which fits snugly into the funnel.

†6. In what characteristic property must two solids differ if they are to be separated merely by being dissolved at room temperature and filtered?

7. Much salt is obtained from salt mines, in which great masses of salt occur mixed with insoluble earthy impurities. What steps can be taken to purify the salt?

8. How could drinking water be obtained from seawater?

5.5 The Separation of a Mixture of Soluble Solids

In the previous experiment, you were able to separate two solids because one of them was soluble in water and the other was not. How can a mixture of two soluble solids be separated? To illustrate the method of separation, consider a mixture of 8.0 g of sodium chloride and 5.0 g of potassium nitrate. At room temperature these two substances have nearly the same solubility (Figure 5.6). Therefore, at room temperature both the 8.0 g of sodium chloride and the 5.0 g of potassium nitrate will dissolve.

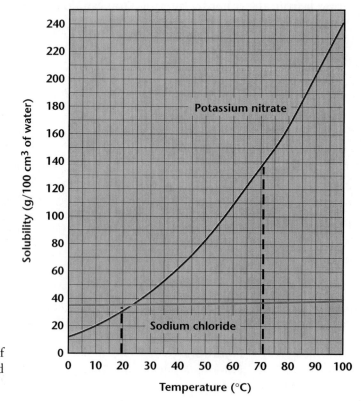

Figure 5.6
Solubility curves of sodium chloride and potassium nitrate.

Suppose we dissolve the mixture in water at 70°C. At this temperature the solubility of potassium nitrate is much greater than that of sodium chloride. If we use just enough water to dissolve the potassium nitrate at 70°C, most of the sodium chloride will remain undissolved and can be filtered out.

After filtering, we let the solution cool down to room temperature. Almost all the sodium chloride will remain in solution, because its solubility hardly changes with temperature. However, the solubility of potassium nitrate is much lower at room temperature than at 70°C. Therefore, as the solution cools, potassium nitrate will precipitate out of the solution and can be filtered out. This process, called *fractional crystallization*, is summarized in Figure 5.7. By repeated fractional crystallizations, even solids

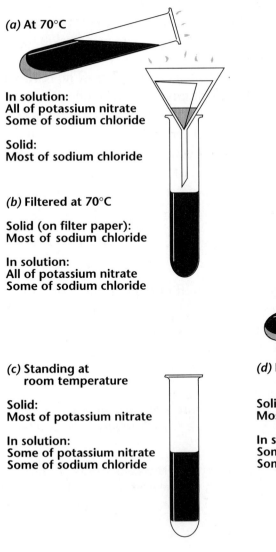

(a) At 70°C

In solution:
All of potassium nitrate
Some of sodium chloride

Solid:
Most of sodium chloride

(b) Filtered at 70°C

Solid (on filter paper):
Most of sodium chloride

In solution:
All of potassium nitrate
Some of sodium chloride

(c) Standing at room temperature

Solid:
Most of potassium nitrate

In solution:
Some of potassium nitrate
Some of sodium chloride

(d) Filtered at room temperature

Solid (on filter paper):
Most of potassium nitrate

In solution:
Some potassium nitrate
Some sodium chloride

Figure 5.7
A diagram of the series of steps for separating a mixture of sodium chloride and potassium nitrate.

whose solubilities are very close to each other at room temperature can be separated.

To achieve a better understanding of the process, we shall review it using the solubility values from Figure 5.6. At 70°C the solubility of potassium nitrate is 138 g/100 cm³. That is, 100 cm³ of water will dissolve 138 g of potassium nitrate. To dissolve 5.0 g of potassium nitrate will require

$$100 \text{ cm}^3 \cdot \frac{5.0 \text{ g}}{138 \text{ g}} = 3.6 \text{ cm}^3.$$

The 5.0 g of potassium nitrate is mixed with 8.0 g of sodium chloride. How much sodium chloride will dissolve in 3.6 cm³ of water? Figure 5.6 tells us that to dissolve 38 g of sodium chloride at 70°C requires 100 cm³ of water. Therefore, in 3.6 cm³ of water only

$$38 \text{ g} \cdot \frac{3.6 \text{ cm}^3}{100 \text{ cm}^3} = 1.4 \text{ g}$$

of sodium chloride will dissolve with the potassium nitrate. The remaining 6.6 g of sodium chloride in the original mixture can be filtered out.

When the filtered solution is cooled down to room temperature, almost all the 1.4 g of sodium chloride will remain dissolved. How much of the 5.0 g of potassium nitrate will precipitate? At 22°C the solubility of potassium nitrate is 37 g/100 cm³. (Again, see Figure 5.6.) The 3.6 cm³ of water in which the 5.0 g of potassium nitrate was originally dissolved at 70°C will now dissolve only

$$37 \text{ g} \cdot \frac{3.6 \text{ cm}^3}{100 \text{ cm}^3} = 1.3 \text{ g}$$

of potassium nitrate, and 3.7 g will be filtered out. Thus, 6.6 g of sodium chloride and 3.7 g of potassium nitrate have been separated from the mixture. Only 1.4 g of sodium chloride and 1.3 g of potassium nitrate remain in the solution. By repeated fractional crystallizations, the two substances can be separated further.

9. Suppose two substances have the same solubility in water at all temperatures. Could you separate them by fractional crystallization? Why or why not?

10. Suppose you wish to separate more of the two solids, sodium chloride and potassium nitrate, remaining in solution, as described at the end of Section 5.5. How would you proceed to do this?

11. Suppose that you planned to separate a mixture of 8 g of sodium chloride and 5 g of potassium nitrate at 80°C rather than at 70°C. How would this higher temperature affect the separation of the two solids in this mixture?

EXPERIMENT
5.6 Paper Chromatography

So far you have learned how differences in density, boiling point, and solubility can be used to separate substances. In this experiment you will investigate a method that works even when the substances in the mixture are present in only minute quantities.

In a plastic bottle containing water, hang a strip of filter paper streaked with green ink from a felt-tip pen, as shown in Figure 5.8. When the color has risen up the paper to about 2 or 3 cm below the top, remove the paper and hang it up to dry.

- How many different substances can you identify?

- Can you put the substances back together again to make green ink?

Figure 5.8
One method of making a paper chromatograph. The point of the filter-paper strip extends into the water so that the ink streak is about 1 cm above the water. The strip is held in place by a stopper.

Cut out each of the colored sections, and put each one in a separate test tube. Add between 0.5 cm³ and 1 cm³ of water to each tube.

- Do the colored substances dissolve?

Pour the liquids from all three test tubes into a single test tube.

- What color is produced?

5.7 A Mixture of Gases: Nitrogen and Oxygen

So far in this chapter you have learned how to separate mixtures of liquids, mixtures of solids and liquids, and mixtures of solids. We have not yet considered the separation of mixtures of gases. If a gas is dissolved in a liquid, all we have to do is heat the mixture. The solubility of gases decreases with temperature (Table 4.2, page 82).

When a glass of cold water warms up to room temperature, bubbles of dissolved air appear on the sides of the glass. If the water is warmed further, more air bubbles appear, and some rise to the top long before the water reaches its boiling point. Separating mixtures of gases alone, however, requires different methods from those used so far.

Mixtures of gases are very common. Table 5.1 lists four gases that are mixed together in the petroleum fraction called "fuel gas."

There are a number of ways to separate gases. One of them, which is widely used, is to cool the mixture until it condenses to form a liquid. Then we can make use of the different boiling points of the various liquids and fractionally distill the cold liquid. The gases are thus collected one by one, as the boiling temperature levels off at new plateaus.

Air that is liquified and fractionally distilled in this way separates mainly into two fractions: A glowing splint bursts into a bright flame when placed in one of them, but even a splint that is burning brightly goes out when placed in the other. Neither gas turns limewater milky. The gas that causes the glowing splint to burst into flame is oxygen, and the one that does not is nitrogen. These two gases together make up about 99 percent of the gases in air. Nitrogen makes up about 80 percent of the atmosphere, and oxygen about 20 percent. The densities, melting points, and boiling points of nitrogen and oxygen are given in Table 5.2.

The cheapest way of obtaining oxygen and nitrogen is to condense air into a liquid and then fractionally distill it. Most of the oxygen and nitrogen commercially manufactured is produced by this method.

12. **How could you separate propane gas from air?**

5.8 Low Temperatures

The melting points and boiling points given in Table 5.2 are far below any temperature you can reach in your laboratory. How is it possible to cool things to such low temperatures?

Table 5.2

Gas	Density (g/cm³)	Melting point (°C)	Boiling point (°C)
Nitrogen	1.2×10^{-3}	−210	−196
Oxygen	1.3×10^{-3}	−218	−183
Hydrogen	8.4×10^{-5}	−259	−253

The densities are given for atmospheric pressure and room temperature.

One method of cooling gases depends on the fact that very highly compressed gases cool when allowed to expand. Figure 5.9 shows how this effect can be used to cool air to temperatures low enough to liquefy it. Air at very high pressure and room temperature flows down the long tube in the center and escapes through a small opening at the bottom. As it escapes, it expands and cools. When the flow is first started, the escaping air does not cool enough to condense into liquid air. However, as this escaping cold air flows up past the long tube, it cools the air moving down inside the tube. Thus, when the air inside escapes at the bottom, it is already cold and cools off still more on expansion. After the apparatus has run for some time, the expanding air cools enough so that some of it condenses into liquid and collects at the bottom of the

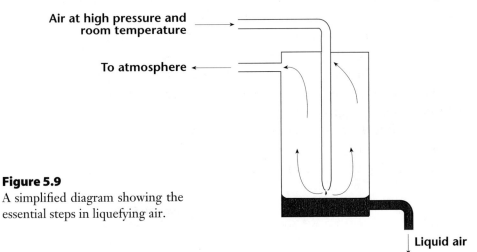

Air at high pressure and room temperature

To atmosphere

Liquid air

Figure 5.9
A simplified diagram showing the essential steps in liquefying air.

apparatus. Of course, the whole apparatus shown in the figure must be well insulated to keep the inside cold. Actual liquid-air machines are more complicated than this simplified diagram shows, but many of them operate on the same principle.

It is one thing to produce very low temperatures, but another to measure them. The reason is simple: all substances that are liquid at room temperature freeze well above the temperatures given in Table 5.2. These temperatures were first measured using a gas thermometer (Figure 5.10). Helium is the most common gas used in gas thermometers because of its low boiling point. A gas thermometer consists of two main parts: a bulb and a barometer connected by a thin tube. The bulb is placed in the liquid or vapor whose temperature is to be measured. The volume of the bulb is much greater than the volume of the thin tube

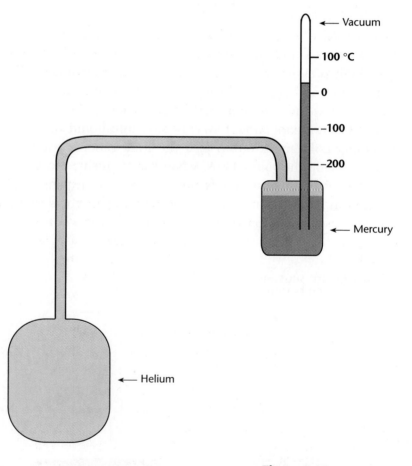

Figure 5.10
A schematic drawing of a gas thermometer.

leading out of it. Therefore, almost all the helium is at the temperature of the liquid or vapor whose temperature is to be measured. In this respect the bulb of a gas thermometer resembles the bulb of a regular thermometer. The pressure of the helium at constant volume (see Section 3.5) depends only on its temperature. Thus, by connecting the bulb to a barometer, the scale on the barometer can be calibrated to read temperature.

13. a. Were all parts of the thermometer that you used in Experiment 5.1, Fractional Distillation, at the same temperature during the experiment?

b. Why could you use the thermometer in the experiment despite your answer to part (a)?

5.9 Mixtures and Pure Substances

In this chapter we have found ways to separate different substances from each other by using characteristic properties. A difference in density can be used to separate two solids; solids can also be separated by differences in solubility. A difference in the boiling points of different liquids enables us to separate them by fractional distillation.

Suppose we experiment with a piece of solid material to see if we can separate it into two or more substances. First we grind it up and mix it with water, stirring it thoroughly. We observe that some particles of a yellowish solid float on the surface, while particles of a gray solid sink to the bottom. We skim off the floating material, whose density is obviously less than that of water. We dry it, call it fraction 1, and set it aside. Then we filter the water and the more dense solid that is in the bottom of the test tube. This solid, which remains on the filter paper, we dry, label fraction 2, and set aside also. We know that these two solids, fractions 1 and 2, are different substances, because they have different densities.

We now test the filtrate to see if any material has dissolved in the water. When we evaporate the water, we find a small amount of white solid. This substance, which is different from both fractions 1 and 2 because it is soluble in water, we call fraction 3. The whole process of the separation of three fractions from a piece of solid material is diagrammed on the next page in Figure 5.11.

We now have separated out three different substances, but perhaps each of these can be further separated. To find out, we use other separation methods. We may, for example, try to melt and even fractionally distill each of the fractions, or we may try dissolving them in different

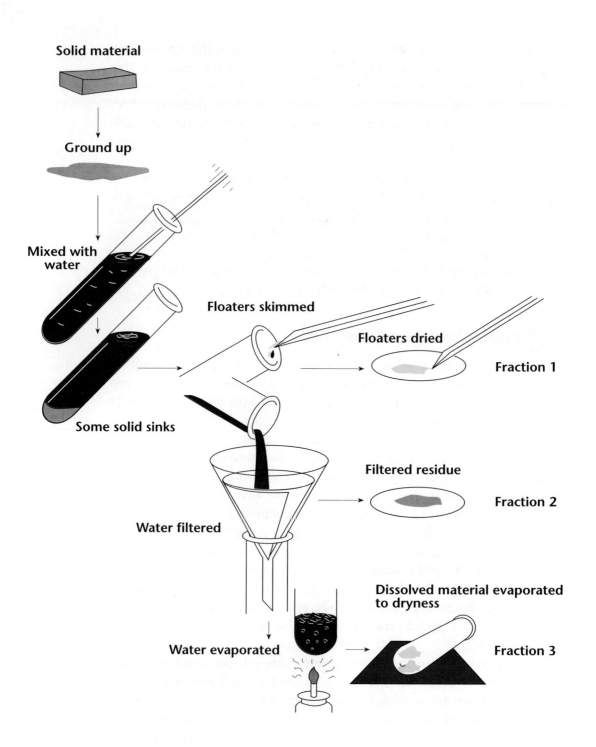

Solid material

Ground up

Mixed with water

Some solid sinks

Floaters skimmed

Floaters dried

Fraction 1

Water filtered

Filtered residue

Fraction 2

Water evaporated

Dissolved material evaporated to dryness

Fraction 3

Figure 5.11
A diagram of an example of the series of steps for separating
three fractions obtained from a ground-up solid.

liquids. Suppose that boiling, melting, and mixing with alcohol and other liquids do not produce anything with characteristic properties different from those of the three original fractions. By using all these various tools of separation again and again, we find that the characteristic properties of our three fractions remain unaltered. We call such substances whose properties are not changed by repeating any of these procedures *pure substances*.

Suppose we mix together all the pure substances that we obtain in this way. If the resulting material has the same characteristic properties as the original sample, we can say that the original sample was a *mixture* of the pure substances. For example, if you mixed all the fractions you separated in Experiment 5.1, Fractional Distillation, you would end up with a substance that has the same properties with which you started. Similarly, mixing together the various fractions obtained from the distillation of liquid air would yield a gas with exactly the same properties as ordinary air. Therefore, air is a mixture.

Note that many properties of a mixture are intermediate between the properties of the pure substances that form the mixture. For example, the density of air is between the density of nitrogen and that of oxygen. A mixture of alcohol and water will smell like alcohol, but it will burn only if it contains enough alcohol.

However, it is not always possible to restore the original sample by mixing the parts separated by the various processes. For example, mixing the distillation products of petroleum will yield a liquid similar but not identical to the original sample.

A simple example of a substance that breaks up when heated and that is not a mixture is mercuric oxide, an orange powder. If you try to determine its melting point, you will discover that when it is heated, mercuric oxide gives off a gas. If you test this gas, you will find it to be oxygen. Furthermore, you can detect some droplets of a silvery liquid—mercury—in the test tube. Mixing the oxygen and the mercury together, however, will not restore the mercuric oxide. The two components remain separated as a gas and a silvery liquid.

Mercuric oxide is a pure substance that cannot be separated into simpler substances by most of the methods we used to separate mixtures. Once mercuric oxide has been separated by heating, it cannot be put back together simply by mixing.

14. **Suppose you mixed together all the fractions from the fractional distillation of the liquid in Experiment 5.1, Fractional Distillation. What do you think would be the properties of this liquid?**

15. A sample of a liquid was boiled for 12 minutes. During that time, the boiling point remained constant, and the volume was reduced to half. Is the liquid a pure substance?

16. In earlier times people would search out sandy stream beds in which small particles of gold were mixed with the sand. They separated this gold from the sand by "panning." Find out how this was done. What characteristic property of the substances made panning possible?

17. Figure 5.2 shows four oil wells drilled into oil-bearing porous rock. Can you suggest some method, other than drilling deeper, for getting more oil from well *D* after the oil level drops below the end of the well?

18. Using the data in Table 5.1 (page 93), draw and label a possible distillation curve for a mixture of hexane, nonane, and tetradecane.

19. Oil and vinegar are often served as salad dressing. They can be poured into the pitcher shown in Figure C. The pitcher has a spout on each side. Yet, oil comes out on one side and vinegar on the other. How do you think the pitcher works?

Figure C
For problem 19

20. a. If a solution containing 40 g of potassium nitrate in 100 cm³ of water at 100°C is cooled to 25°C, how much potassium nitrate will precipitate out of solution? (See Figure 5.6.)

    b. Suppose that the 40 g of potassium nitrate is dissolved in only 50 cm³ of water at 100°C. How much potassium nitrate will precipitate out if the solution is cooled to 25°C?

21. Suppose you dissolve 30 g of sodium chloride in 100 cm³ of water at 100°C and boil away 50 cm³ of the water.

    a. How many grams of sodium chloride will remain in solution?

    b. How many grams will precipitate out of solution?

22. Suppose you dissolve 40 g of potassium nitrate in 100 cm³ of water at 100°C.

    a. If half the solution is poured out, how many grams of potassium nitrate will the remaining solution contain?

    b. Now, instead of pouring out part of the solution, you boil away 50 cm³ of water. How many grams of potassium nitrate will remain in solution at 100°C?

    c. If the solution remaining in (b) were cooled to 25°C, how much potassium nitrate would precipitate out of solution?

23. When ethanol is to be used for industrial or commercial purposes other than as a beverage, it is customarily denatured. What, in general, do you think some of the properties of the added substance might be?

24. How would you separate a mixture of powdered sugar and powdered citric acid?

25. The mineral called "Gay-Lussite" appears to be a pure substance, but it is actually a mixture composed of calcium carbonate (limestone), sodium carbonate (soda ash), and water. Describe how you would go about separating these three substances from the rock. Some of the properties of calcium carbonate and sodium carbonate are listed in the following table.

Property	Calcium carbonate	Sodium carbonate
Melting point	Decomposes at 825°C	851°C
Solubility in alcohol	Insoluble	Insoluble
Solubility in hydrochloric acid	Soluble	Soluble
Solubility in water	Insoluble	7 g/100 cm³ at 0°C; 45 g/100 cm³ at 100°C

26. You can use paper chromatography to separate the components in many common substances. Try this technique with any of the following substances you can find at home: tomato paste, different colors and brands of ink, the coloring in leaves and vegetables (grind the leaves first and dissolve them in alcohol), and flower petals.

27. Chlorophyll can be extracted from leaves by grinding them with alcohol to give a dark-green solution. By careful application of paper chromatography, bands of yellow and red color, as well as green bands, can be detected. What other reason do you have to suspect the presence of substances producing these colors in leaves? Why do you ordinarily not see them?

28. As liquid air boils away, the remaining liquid becomes richer in one of the two gases—nitrogen or oxygen. Which one is it? How do you know?

29. What would you do to separate
    a. alcohol from water?
    b. sodium chloride from sodium nitrate?
    c. nitrogen from oxygen?

THEME FOR A SHORT ESSAY

You devoted several hours to the "sludge test." Do you think this test is a good way to assess your progress in this course? Why, or why not? Include enough information about the test so that a friend who has never heard about the "sludge test" will be able to understand your reasoning.

Chapter 6
Compounds and Elements

6.1 Breaking Down Pure Substances

At the end of the last chapter, you learned about a pure substance, mercuric oxide, that can be broken down into two different pure substances—mercury and oxygen—by heating. The properties of these two components are quite different from those of mercuric oxide. Furthermore, the mercury and oxygen cannot be put back together to form mercuric oxide simply by mixing.

Similarly, baking soda, which you heated at the beginning of the course, is also a pure substance. It decomposed into a solid, a liquid, and a gas. If you had tested these remaining substances, you would have found them also to be pure substances. Mixing them will not bring back the baking soda.

Heating is not the only method for decomposing a pure substance. In the next section, you will use a different method.

EXPERIMENT
6.2 The Decomposition of Water

Water is one of the pure substances you have separated out of many mixtures. But water itself cannot be separated into other substances by the methods used in the previous chapter. After heating, distilling, and freezing water, we always end up with plain water. To separate water into simpler substances, we shall need a new method, different from any used so far. Such a method, developed at the beginning of the nineteenth century, relies on electricity and is called *electrolysis*.

CAUTION: Be sure to wear safety glasses.

Set up the apparatus shown in Figure 6.1. Connect the wires to the battery and see if anything happens. Since this reaction is very slow when pure water is used, something must be added to speed up the process. Try adding from 10 to 30 cm^3 of sodium carbonate solution to the water.

Disconnect the battery when one of the test tubes is nearly full of gas. Mark the volume of gas in each tube with a grease pencil or rubber band.

You have probably heard that water is made up of hydrogen and oxygen. Test the gases to see if you come to the same conclusion using burning and glowing splints. Be sure to remove the test tubes from the water in such a way that you do not lose any gas.

Measure the volumes of the gases in the test tubes. To compare these two volumes, divide the volume of the hydrogen by the volume of the

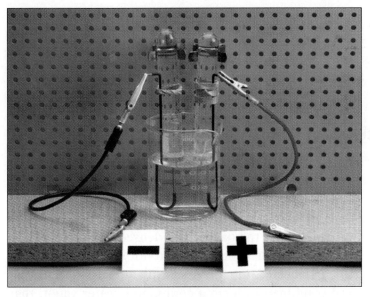

Figure 6.1
Apparatus for decomposing water. The electrodes are clamped alongside the inverted test tubes and connected to a battery (not shown). Any gas that forms on either of the stainless-steel electrode tips will be collected when it rises and displaces water from its test tube. No reaction is observed until a small amount of sodium carbonate solution is added to the water.

oxygen. This gives you the ratio of the volume of hydrogen to the volume of oxygen. Compare your ratio with the ratios obtained by students who used different amounts of sodium carbonate solution mixed with the water.

- Does the amount of solution added to the water affect the volume ratio?

- What is the mass ratio of hydrogen to oxygen?

A pure substance that can be broken up into two or more pure substances is called a *compound*. The characteristic properties of a compound are quite different from those of its components.

1. In decomposing water, suppose you had filled many test tubes with gas. As the water disappeared from the beaker, suppose you refilled it without adding more sodium carbonate. You always would have found the ratio of hydrogen to oxygen produced to be constant. What does this tell you about the source of the gases? About the sodium carbonate?

†2. a. What is the total mass of oxygen and hydrogen that can be produced by the decomposition of 180 g of water by electrolysis?

   b. If all the hydrogen produced were burned in the air to form water, what mass of water would result?

†3. Two test tubes contain equal volumes of gas at atmospheric pressure. If one contains oxygen and the other helium, is the mass of gas in both tubes the same?

4. From your data in Experiment 6.2, The Decomposition of Water, and Table 3.1 calculate the ratio of the mass of oxygen produced to the mass of hydrogen produced. How does your ratio compare with those found by other members of your class?

6.3 The Synthesis of Water

In the last experiment, you could have mixed oxygen and hydrogen gas together in a test tube. Nothing would have happened unless you had ignited the mixture. Then a violent reaction would have occurred. With the proper equipment, the gases could have been separated before ignition by cooling them down until the oxygen condensed at −183°C, leaving the hydrogen as a gas. But this method would not have worked if the mixture of gases had first been ignited. As a result of the ignition, the two gases would have combined to form a compound, water vapor, that cannot be separated into hydrogen and oxygen by condensation. The combining of substances to form a compound is called *synthesis*. This process is the opposite of decomposition.

When you electrolyzed water, you found the ratio of the volume of hydrogen to the volume of oxygen produced. No matter how much water you decomposed, the ratio remained the same. Of course, you might have expected this, since all the water you used came from the same source. But can hydrogen and oxygen combine in different proportions, or will they combine only in the same volume ratio?

To answer this question, we could try adding different volumes of hydrogen to a fixed volume of oxygen, igniting each mixture, and measuring how much of each gas, if any, remains uncombined. We have done this experiment. First we filled a special shock-resistant glass tube with water and inverted it in a tray of water. A volume V of oxygen was then bubbled into the bottom of the tube. The oxygen rose to the top (The water levels are marked by rubber bands.), as shown in the photograph labeled "Step 1" in Figure 6.2(a).

We then added a volume V of hydrogen to the same tube ("Step 2"). Next, we ignited the mixture of gases by sending an electric spark through the mixture. After ignition, we measured the volume of gas that remained unreacted ("Result").

In Figure 6.2(a), the volume of hydrogen added was equal to the volume of oxygen. After ignition, a volume of gas equal to $1/2V$, or half the volume of oxygen, had not reacted.

In Figure 6.2(b), the volume of hydrogen used was $2V$, or twice the volume of oxygen. In this case, practically all of the mixture reacted.

In Figure 6.2(c), the volume of hydrogen added before ignition was $3V$. A volume equal to $1V$ of gas did not react.

Figure 6.2
In each of the three parts of this figure, (a), (b), and (c), the tube on the left was first full of water (not shown). Then, a volume V of oxygen was bubbled into the bottom of the tube.

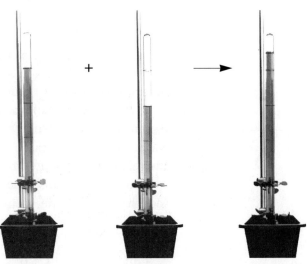

(a) An equal volume of hydrogen was added to the oxygen, and the mixture was ignited by an electric spark. Note that about $\frac{1}{2}V$ of gas remained unreacted.

1V of oxygen   **1V of hydrogen**   **Result: $\frac{1}{2}V$ of gas left**

(b) A volume $2V$ of hydrogen was added. All the gas reacted. (The tiny bubble is within the uncertainty of the volume measurements.)

1V of oxygen   **2V of hydrogen**   **Result: no gas left**

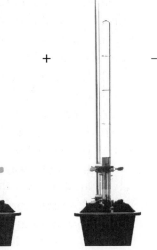

(c) A volume $3V$ of hydrogen was added to the oxygen. Now a volume equal to $1V$ remained unreacted.

1V of oxygen   **3V of hydrogen**   **Result: 1V of gas left**

From Figure 6.2, you can see that the ratios of the volumes of hydrogen to oxygen before ignition in (a), (b), and (c) were 1/1, 2/1, and 3/1, respectively. When tested after ignition, the gas remaining in (a) was oxygen and that in (c) was hydrogen. (The volume of the bubble remaining in (b) was too small to identify the gas.) The increased amount of water in the tubes came from the tray of water. The water produced in the reactions amounted to only a few drops.

The results of this experiment are summarized in Table 6.1. As you can see, the ratio of the volume of hydrogen to the volume of oxygen that combined to form water was the same in each case, regardless of the ratio of the volumes of the two gases in the mixture before they were ignited.

This ratio is the same as the ratio of hydrogen gas to oxygen gas obtained by the careful electrolysis of water in Experiment 6.2, The Decomposition of Water. The experiment in Figure 6.2 shows what happens to any excess gas. Over the range of volumes used, if the volume of either gas was greater than that needed for a hydrogen-to-oxygen ratio of 2:1, only part of the gas reacted. The excess remained uncombined.

Table 6.1

Tube	Initial volume of oxygen (cm³)	Initial volume of hydrogen (cm³)	Final volume of gas (cm³)	Volume of oxygen that combined (cm³)	Volume of hydrogen that combined (cm³)	Ratio hydrogen to oxygen
1	25.0	25.0	12.4 oxygen	12.6	25.0	1.98
2	25.0	50.0	0.8	25.0	49.2	1.97
3	25.0	75.0	24.6 hydrogen	25.0	50.4	2.02

†5. If 18 g of water is decomposed into hydrogen and oxygen by electrolysis, 16 g of oxygen and 2 g of hydrogen are produced. Using the table of densities in Chapter 3 (page 58), find (a) the volume of water decomposed and (b) the volume of hydrogen produced.

6. Suppose you were not given the information about the identity of the remaining gases in Figure 6.2. Suggest a line of reasoning that would lead to the conclusion that the remaining gas in (a) is oxygen, and in (c) is hydrogen.

7. Suppose you mixed 100 cm³ of oxygen with 200 cm³ of hydrogen. The volumes of both gases are measured at atmospheric pressure and at room temperature.
   a. Calculate the mass of oxygen and the mass of hydrogen used.
   b. If you ignited the mixture, what mass of water would result from the reaction?
   c. What volume of water would be produced?

8. If in problem 7 you had used 100 cm³ of oxygen but only 50 cm³ of hydrogen, what mass of water would have resulted?

†9. Three tubes are filled with a mixture of hydrogen and oxygen in a manner similar to that used in Figure 6.2. If the three tubes contain the following volumes of hydrogen and oxygen, what is the volume of the unreacted gas remaining in each tube after ignition

Tube	Volume of oxygen (cm³)	Volume of hydrogen (cm³)
I	25	75
II	50	50
III	25	50

EXPERIMENT
6.4 The Synthesis of Zinc Chloride

You have learned that hydrogen and oxygen combine in a definite ratio, no matter how much of each mix together and ignite. That reaction involves two gases. You will now investigate what happens when a metal is dissolved in an acid: in this case, zinc in hydrochloric acid.

Everyone in your class will use the same amount of hydrochloric acid, but different groups will add different amounts of zinc. When the reaction is complete, you will determine the mass of zinc that reacted. You

Figure 6.3
Zinc is dissolved in 10.0 cm³ of hydrochloric acid in a test tube, which is placed in a beaker of cold water to keep the solution from getting too hot.

will then evaporate the remaining liquid and mass the solid residue. Then each group will calculate the ratio of the mass of zinc reacted to the mass of solid product formed (the solid is called *zinc chloride*).

The reaction between the zinc and the acid will produce much heat. To keep the mixture cool, you can perform the reaction in a large test tube placed in a beaker of cold water. (Figure 6.3)

Mass an exact amount of zinc between 0.5 g and 4 g. Place the zinc in a test tube and add 10.0 cm^3 of hydrochloric acid.

CAUTION: Always wear safety glasses when you work with acids. Be careful not to spill any of the solution on your books or clothes. If some acid spills on your hands, wash them thoroughly with water.

- What is the gas given off in the reaction?

The reaction at first is quite vigorous. However, to make sure that it is complete, allow the mixture of zinc and acid to stand overnight. In the next period, pour the liquid from the test tube into an evaporating dish that you have already massed. If there is still zinc left over from the reaction, be sure the solid stays in the test tube when you pour the solution. Wash the test tube, and any zinc remaining, with 5 cm^3 of water.

Figure 6.4
Evaporating a solution in an evaporating dish heated directly over an alcohol burner or a microburner.

- Why should you add the washing water to the evaporating dish?

Dry the leftover zinc, and mass it.

- How much of the metal reacted with the hydrochloric acid?

The evaporating dish can be placed on a burner stand and heated, as shown in Figure 6.4. If the solution begins to spatter, gently move the flame back and forth.

Heat the material until it appears to be dry. Continue heating until the solid begins to melt and a tiny pool of liquid forms in the bottom of the evaporating dish. You can mass the dish and its contents as soon as they have cooled.

- What is the ratio of the mass of zinc reacted to the mass of zinc chloride formed?

- Compare your results with those of your classmates. Did an excess of either zinc or hydrochloric acid affect your results?

- If the zinc chloride were not completely dry when massed, how would this affect your ratio?

10. Suppose that in the synthesis of zinc chloride, you dissolved 5 g of zinc.
    a. How much product would you get?
    b. What would be the ratio of zinc to the product?
    c. What would be the ratio if you dissolved 50 g of zinc?

11. In a certain package of seed corn, the number of red seeds was 36 and the number of yellow seeds was 24. In a second package, the number of red seeds was 51 and the number of yellow seeds was 34.
    a. What is the ratio of the number of red seeds to the number of yellow seeds in each package?
    b. What is the ratio of the number of red seeds to the total number of seeds in each package?

12. Suppose that there are 10 boys and 15 girls in a class.
    a. What is the ratio of boys to girls in the class?
    b. What is the ratio of boys to the total number of students in the class?
    c. What would be the ratio of boys to girls if the class were three times larger but the ratio of boys to the total number of students were the same?

†13. When various amounts of zinc react with hydrochloric acid, zinc chloride and hydrogen are produced. Which of the following ratios between the masses of various products and reacting

substances are constant regardless of the amounts of zinc and acid mixed together?

a. $\dfrac{\text{Zinc added}}{\text{Zinc chloride produced}}$ b. $\dfrac{\text{Zinc used up}}{\text{Hydrochloric acid used up}}$

c. $\dfrac{\text{Zinc used up}}{\text{Hydrogen produced}}$ d. $\dfrac{\text{Zinc used up}}{\text{Zinc chloride produced}}$

e. $\dfrac{\text{Zinc added}}{\text{Hydrochloric acid added}}$ f. $\dfrac{\text{Zinc chloride produced}}{\text{Zinc used up}}$

g. $\dfrac{\text{Hydrochloric acid used up}}{\text{Hydrogen produced}}$

6.5 The Law of Constant Proportions

In the last two sections, you studied the synthesis of two compounds: water from oxygen and hydrogen, and zinc chloride from zinc and hydrochloric acid. You found that hydrogen and oxygen combine only in the definite mass ratio of 0.13. It does not matter in what proportion these gases are mixed. When they are ignited, the hydrogen and oxygen that react do so in a definite proportion to produce water. In the last experiment, zinc chloride was produced when zinc combined with the chlorine in the hydrochloric acid. The ratio of the mass of zinc that reacted to the mass of zinc chloride was constant. It was independent of an excess either of hydrochloric acid or of zinc. This means that the ratio of zinc to the chlorine with which it combined also was constant.

We consider both water and zinc chloride to be compounds and not mixtures because the characteristic properties of each are quite different from those of the substances from which it is made. Do all substances combine in a constant proportion when they form compounds?

When you investigated the law of conservation of mass, there were two experiments in which compounds were formed. In one of these (Experiment 2.5), you heated copper and sulfur to make a new substance. Suppose we repeat the experiment, keeping the mass of copper constant and varying the mass of sulfur. Will we find that the mass of copper that reacts remains in a constant ratio to the mass of the product, independent of how much sulfur we use?

This experiment has been done many times. As long as there is more than enough copper to react with all the sulfur, the ratio of the mass of the copper that undergoes reaction to the mass of the product remains fixed. But with an excess of sulfur, the ratio decreases. In this case it appears at first that when copper and sulfur combine to form a compound, the ratio of the masses that react can vary.

The early chemists strongly disagreed about the relative amounts of substances that react to form compounds. On one side was a distinguished French chemist, Claude Louis Berthollet (1748–1822). He claimed, on the basis of experiments like the one with copper and sulfur, that a pair of substances can combine in any proportion to form a compound. On the other side was another distinguished French chemist, Joseph Louis Proust (1754–1826). He based his answer on evidence obtained from experiments that showed constant proportion, like the synthesis of water and of zinc chloride.

Proust suggested a new law of nature, *the law of constant proportions*, which he stated thus in 1799: "We must recognize an invisible hand which holds the balance in the formation of compounds. A compound is a substance to which Nature assigns fixed ratios; it is, in short, a being which Nature never creates other than balance in hand." In plainer language, the law that Proust formulated can be stated as follows: When two substances combine to form a compound, they combine in a constant proportion. The ratio of the masses that react remains constant, no matter in what proportions the substances are mixed. If there is too much of one of the substances in the mixture, some of it just will not react.

When the law of constant proportions was formulated, the evidence in its favor was much weaker than the evidence you gathered for the law of conservation of mass at the end of Chapter 2. In spite of much evidence that supported Berthollet's stand and that Proust could not explain, Proust was confident enough to claim constant proportions as a law of nature.

The explanation of results like those in the copper/sulfur reaction that seemed to support Berthollet's stand will be discussed later in the course.

14. If you make a solution of salt and water, over what range of values can you vary the mass ratio of salt to water at a given temperature?

15. a. Do you think gasoline is a single compound? See Table 5.1 (p. 93).

    b. Would you expect gasoline from different pumps to be the same? Why?

EXPERIMENT
6.6 A Reaction with Copper

Some substances react very fast. As you learned in Section 6.3, when a test tube of hydrogen and oxygen is ignited, the reaction is very fast. Indeed, it is explosive and ends in a fraction of a second. Solids usually do not react as fast as gases. In this experiment you will investigate how the reaction of finely ground copper with oxygen in the air proceeds with time.

Mass a dry crucible and add about 1 g of copper dust. Now find, as accurately as possible, the total mass of the copper and the crucible.

Heat the copper, as shown in Figure 6.5, for 2 minutes. While heating, watch the copper carefully.

CAUTION: Always wear safety glasses when you use a burner.

* Does a reaction take place?

When the crucible is cool, mass the crucible and its contents.

* Did the crucible and its contents gain or lose mass?
* Is your answer to the previous question evidence that a reaction has taken place?
  * What do you predict will happen if you continue to heat the crucible for an additional 10 to 15 minutes? Try it.

  Break up the contents of the crucible with a scoopula and examine the pieces.
* Do you think that all the copper has reacted? If not, what fraction do you estimate did not react?

NOTE: Do not discard the black solid in your crucible. Pour the pieces of solid into a test tube and save them for use in the next experiment.

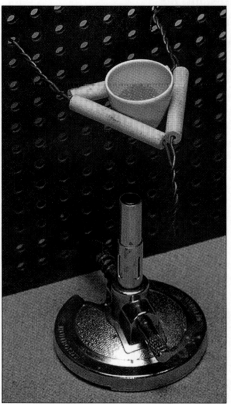

Figure 6.5
Powdered copper in a crucible supported by a triangle over a microburner. The wires of the triangle are bent so that they can be hooked into holes in the pegboard for rigid support.

EXPERIMENT
6.7 The Separation of a Mixture of Copper Oxide and Copper

Suppose all the copper that you started with in the preceding experiment had reacted with oxygen. In that case the black solid present at the end of the experiment would be the pure substance copper oxide. If, however, not all the copper reacted, then the black solid would be a mixture of copper and copper oxide. In order to determine whether the black solid is a mixture, you will attempt to separate it.

A good way to separate a mixture like copper oxide and copper is to place the mixture in a solvent that will dissolve one of the components but not the other. Copper will not easily dissolve in hydrochloric acid, as your teacher can show you. Copper oxide, on the other hand, is soluble in hydrochloric acid.

Place in a test tube all the black solid you obtained in the previous experiment. Add 5 cm^3 of hydrochloric acid and stir gently with a glass stirring rod for 5 to 10 minutes.

CAUTION: Always wear safety glasses when you work with acids. Be careful not to get any hydrochloric acid on yourself or your books or clothes. If some acid spills on your hands, wash them thoroughly with water.

After letting the solid settle to the bottom, slowly pour off the liquid into another test tube. Now wash the remaining solid several times with water and discard the washings.

- What does the remaining solid look like?
- Is the black substance a mixture?

NOTE: Do not discard the acid solution. You will need it for the next experiment.

6.8 Complete and Incomplete Reactions

Many reactions, like the reaction of copper with oxygen, are slow. It is difficult in such cases to tell when one of the reacting substances has been completely used up. Because the copper in your crucible changed to a black solid, you may have assumed that *all* the copper had reacted. This would have been an incorrect assumption, as the presence of copper in the black substance has shown.

Reactions like the reaction of copper with oxygen are called *incomplete reactions*, because neither one of the reacting substances, copper or oxygen, is completely used up during the reaction. Other reactions, such as the reaction of zinc with hydrochloric acid, which you studied in Section 6.4, are *complete reactions*. Zinc will continue to react with hydrochloric acid until either one of the two has been completely used up.

Recall the controversy that existed among early chemists about the validity of the law of constant proportions (Section 6.5). At least part of the controversy could have been avoided if they had more carefully analyzed their experiments. Some of their disagreements occurred because they did not understand the difference between complete and incomplete reactions. Specifically, they often mistook the masses they mixed for the masses that reacted. As your own experience shows, this mistake is easy to make.

EXPERIMENT
6.9 Precipitating Copper

When you heated a sample of finely ground copper in Experiment 6.6, A Reaction with Copper, you observed an increase in mass. This indicated that something was being added to the copper. A new substance, copper oxide, was formed. In Experiment 6.7, The Separation of a Mixture of Copper Oxide and Copper, you dissolved the copper oxide in hydrochloric acid, and from the black powder you produced a blue-green solution.

What happened to the copper in all these reactions? Did it disappear forever, or is it still in the solution in some form? Putting a piece of zinc in the solution will provide you with an answer. As you already know, zinc in hydrochloric acid will generate heat.

CAUTION: Always wear safety glasses when you work with acids. Put the test tube containing the solution in a beaker half-filled with water.

Put a piece of zinc in the solution you saved from Experiment 6.7, and observe the reaction. While the reaction is going on, break up the solid from time to time with a stirring rod.

After the zinc has finished reacting, pour off the solution from the remaining solid and wash the solid several times with water to clean off the acid. Remove the washed solid from the test tube and dry it quickly by pressing it between two layers of paper towel.

Press a scoopula firmly against the paper towel and move it across the dry solid.

- Can you identify the solid?

---

16. **What would happen if you heated the solid that you recovered in this experiment?**

17. **How do you expect the total mass of solid you recovered in Experiments 6.7 and 6.9 to compare with the initial mass of copper you heated in Experiment 6.6?**

6.10 Elements

Heating copper in air is not the only way to produce a reaction with copper. There are many other reactions, and they all have one feature in common: The product of the reaction always has a mass larger than the mass of the copper. Neither heating nor electrolysis will change copper into something that has a smaller mass than did the copper with which we started.

Contrast this behavior of copper with that of mercuric oxide (Section 5.9). When you heat mercuric oxide, oxygen is given off, and the resulting mercury has a smaller mass than the original quantity of mercury. We reason, therefore, that mercury is a simpler substance than mercuric oxide.

In Experiment 6.2, The Decomposition of Water, you electrolyzed water, causing it to decompose into hydrogen and oxygen. In the same way, molten sodium chloride can be decomposed by electrolysis into chlorine and sodium. The mass of either the chlorine alone or the sodium alone is less than the mass of the sodium chloride from which they were produced. Thus, sodium and chlorine must be simpler substances than sodium chloride. However, just as with copper, none of the methods that can be used to break up other substances work for either sodium or chlorine. Pure substances that do not break up by heating, electrolysis, reacting with acids, or similar methods are called *elements*.

It is not necessary, of course, to try to break up the many thousands of pure substances in order to find out which of them are elements. In many cases we know substances are compounds simply because they can be made by combining other substances. For example, we can "burn" sodium in an atmosphere of chlorine gas. The resulting white solid has the characteristic properties of table salt. Thus, table salt is not an element. It is the compound sodium chloride.

Early chemists using this reasoning were able to identify as possible elements a relatively small number of the pure substances they knew. One such list, proposed by the French chemist Lavoisier in 1789, was accompanied by the following explanation: "Since we have not hitherto discovered the means of separating them, they act with regard to us as simple substances, and we ought never to suppose them compounded until experiment and observation have proved them to be so."

Lavoisier's list of elements included several well-known substances that are also found on today's list of elements. Among them are iron, copper, silver, gold, hydrogen, oxygen, and carbon. From a historical point of view, two substances are of special interest. One was on Lavoisier's list and one was not.

The first substance is lime, which can be produced by strongly heating limestone. Lime was known to the Romans as early as 200 B.C. Many attempts were made over the years to decompose lime: The substance was heated in air, heated in a vacuum, and heated with carbon. Every attempt failed, and most people agreed that lime must be an element. There were a few doubters, who thought that lime was probably a compound of a metal with oxygen. They thought that lime could not be decomposed simply because the usual agent, carbon, which worked with copper oxide and other oxides, was somehow not "powerful" enough to separate the oxygen from the metal.

As so often happens, a new discovery had to wait for a new experimental method. In this case, the electric battery, invented by Alessandro Volta in 1800, was the necessary tool. In 1807, the English chemist Humphry Davy used an electric battery to try to decompose metal oxides. He had already used a battery to decompose molten potash and molten soda to obtain the new elements potassium and sodium. A similar procedure failed to work with lime because lime could not be melted. After much experimentation, however, Davy finally succeeded. He was able, by carefully electrolyzing moist lime, to produce tiny amounts of a new element he called *calcium*.

Missing from Lavoisier's list was a very reactive gas, slightly green in color. It was discovered in 1774 as a by-product of some experiments with muriatic acid. (Muriatic acid we now call hydrochloric acid.) Various people experimented with the new gas and found it similar in many ways to the gases from other strong acids. Acids were considered in those days to be substances that were strongly corrosive and had a sour taste. Since oxygen was believed at that time to be the active ingredient in all acids, this new substance was named "oxymuriatic acid gas."

This name stuck for more than thirty years, mainly because it had been suggested by some very important and respected chemists. Several people tried, unsuccessfully, to decompose the gas. They always blamed their failures on poor methods and poor tools. In 1810 Humphry Davy, after working with his usual patience and brilliance for a full two years, finally announced the results of his long series of experiments. These experiments, Davy wrote, "...incline me to believe that the body improperly called oxymuriatic acid gas has not as yet been decompounded; but that it is a peculiar substance, elementary as far as our knowledge extends, and similar in many of its properties to oxygen gas..." In order that people would no longer be misled into thinking it a compound, he suggested a new name, "based upon one of its obvious and characteristic properties—its color," and called it *chlorine*. Davy's conclusion has stood the tests both of time and new techniques, and chlorine is included in the present-day list of elements.

18. How do you know that water, zinc chloride, and baking soda are not elements?

19. Hydrogen chloride—a gaseous pure substance—can be decomposed into two different gases, each of which acts like a pure substance. On the basis of this evidence alone:
    a. Can hydrogen chloride be an element?
    b. Can either of the other two gases be an element?
    c. Can you be sure that any of the pure substances mentioned is an element?

20. While on a class field trip, a student found a shiny rock that appeared to be a metal. When she returned to the classroom, she heated the rock for a while and found that it lost mass. Further heating did not affect the mass. Could this rock be an element? Explain your answer.

6.11 Elements Near the Surface of the Earth

Of the 109 elements known today, only about 50 are commonly used; they are listed in Table 6.2 on the next page. The elements on the left side of the table are listed in decreasing order of abundance near the surface of the earth. The elements on the right side of the table are listed alphabetically; their abundance is less than that of argon and is not listed.

The part of the earth for which abundance was calculated includes the atmosphere, the oceans, and the ground to a depth of one kilometer below the surface (including that below the oceans). The data used in the calculations are based on the analysis of thousands of samples of soil, sea water, and air collected at many locations.

Note how few elements account for most of the mass around us. The first five elements (oxygen to iron) in the left-hand column of Table 6.2 account for almost 92 percent of the total mass of the elements found near the surface of the earth. Oxygen, the most abundant element,

Table 6.2 Elements Found Near the Surface of the Earth

Name	Symbol	Abundance (in % by mass)	Name	Symbol
Oxygen	O	65.5	Arsenic	As
Silicon	Si	13.7	Boron	B
Hydrogen	H	5.4	Bromine	Br
Aluminum	Al	4.2	Cadmium	Cd
Iron	Fe	3.1	Cesium	Cs
Calcium	Ca	2.4	Cobalt	Co
Sodium	Na	1.7	Germanium	Ge
Magnesium	Mg	1.5	Gold	Au
Chlorine	Cl	0.95	Helium	He
Potassium	K	0.94	Iodine	I
Titanium	Ti	0.32	Krypton	Kr
Nitrogen	N	0.14	Lead	Pb
Sulfur	S	0.061	Lithium	Li
Phosphorus	P	0.060	Mercury	Hg
Manganese	Mn	0.053	Neon	Ne
Fluorine	F	0.028	Osmium	Os
Barium	Ba	0.020	Platinum	Pt
Strontium	Sr	0.019	Polonium	Po
Carbon	C	0.010	Radium	Ra
Vanadium	V	0.0067	Radon	Rn
Chromium	Cr	0.0060	Silver	Ag
Nickel	Ni	0.0050	Tin	Sn
Zinc	Zn	0.0039	Tungsten	W
Copper	Cu	0.0034	Uranium	U
Argon	Ar	0.0023	Xenon	Xe

occurs in many compounds found in the ground. Ordinary sand contains mostly silicon dioxide, a compound of oxygen and silicon. Silicon dioxide is also the main ingredient of glass. Hydrogen, the third element in Table 6.2, is found primarily in the oceans. The 59 elements not listed in Table 6.2 together account for less than 0.001 percent of the mass near the earth's surface.

21. How many elements collectively compose 99 percent of the mass of the earth *accessible* to people?

22. Data for the elemental composition of living matter were not used to calculate the abundance of the elements that are listed in Table 6.2. Even so, the data in Table 6.2 are correct. How do you think that this can be so?

23. The total mass of the one-kilometer layer of ground, the oceans, and the atmosphere is 2.8×10^{21} kg. What is the mass of (a) iron, (b) sulfur, and (c) copper near the surface of the earth?

6.12 The Production of Iron and Aluminum

Our industrial civilization requires large quantities of metallic elements. Among the most important ones are iron and aluminum. Materials found in the ground from which metals can be profitably extracted are called *ores*. Ores of a given element may include a number of different compounds of that element. The most important ores of iron and aluminum are their oxides. Ores are usually found mixed with unwanted substances from which they have to be separated before the metal can be extracted.

It has been known for over three thousand years that if iron ore is mixed with coal or charcoal and heated strongly, iron is produced. In recent times, the same process has been carried out in huge blast furnaces.

Figure 6.6 on the next page provides an overall view of a modern blast furnace. The covered passage on the right contains the conveyor, which carries a mixture of iron ore, coke (a form of carbon), and crushed limestone to the top of the furnace. Air preheated in the four stoves on the left is forced into the lower part of the furnace.

Fierce burning of the coke raises the temperature to over 1,650°C. Carbon monoxide is formed, which, together with the coke, removes the oxygen from the iron. The iron melts and settles to the bottom of the furnace.

Figure 6.6
A general view of a modern blast furnace close to 100 m high. The passageway on the right contains the conveyer belt that brings the ore, coke, and limestone to the top of the furnace. (*Bethlehem Steel Corporation*)

The limestone combines with the impurities in the ore to form slag, which melts and floats on top of the liquid iron. The slag prevents the iron from recombining with the oxygen in the incoming air. The molten iron and slag are drawn off from time to time.

The gases from the top of the furnace contain carbon monoxide. These gases are piped to the stoves, and the carbon monoxide is burned there to preheat the air going into the furnace. Figure 6.7 is a schematic drawing of the blast furnace and one of the stoves. Figure 6.8 shows molten iron flowing through clay-lined channels before being further processed into steel.

Figure 6.7
A schematic drawing of the blast furnace and stove. Gases from the blast furnace are mixed with air and burned in the right side of the stove. The hot gases then go through a checkerwork of fire bricks, making the bricks red hot. After about an hour, the burning is stopped and fresh air is pushed through the checkerwork in the opposite direction. The air heats up and is forced into the blast furnace.

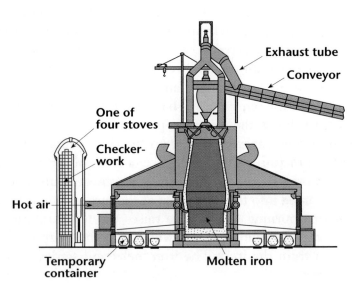

Figure 6.8
Molten iron flowing from the furnace. Because of its high
temperature it glows brightly. (*Bethlehem Steel Corporation*)

Aluminum, like iron, is a widely used element. The density of aluminum is only about one-third that of iron. Therefore, aluminum is used where low mass is desirable, as in airplane bodies. The most important ore of aluminum, bauxite, is about 50 percent aluminum oxide. Pure aluminum oxide is obtained in a series of steps, including dissolving, filtering out the unwanted solids, and crystallizing the aluminum oxide.

Unlike iron, aluminum cannot be freed from its oxide in a blast furnace. Only about 100 years ago was a way discovered to separate aluminum from oxygen by electrolysis. Aluminum oxide is first dissolved in molten cryolite (a compound of sodium, aluminum, and fluorine) at a temperature of around 1,000°C. The electrolysis takes place in a large rectangular metal container called a cell. A schematic drawing of a cell is shown in Figure 6.9 on the next page. The cell is lined with carbon, which serves as the negative electrode (−). A large carbon block at the center forms the positive electrode (+). The electric current reaches the block through a set of pins. The cryolite is heated and

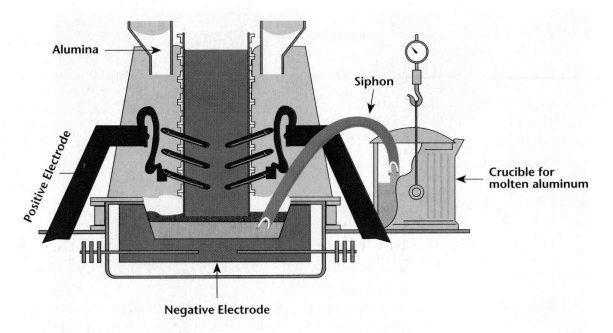

Alumina

Siphon

Positive Electrode

Crucible for
molten aluminum

Negative Electrode

Figure 6.9
A schematic drawing of an aluminum-producing cell.
(*Courtesy Reynolds Metals Company, Richmond, VA*)

kept molten by the passage of electricity through it. Liquid aluminum gathers at the bottom of the cell and is siphoned off. The oxygen from the aluminum oxide combines with the carbon in the block. The resulting gases are vented off, and the carbon block is slowly eaten away. The block is pushed down and from time to time a new set of pins is connected to the main conductor.

FOR REVIEW, APPLICATIONS, AND EXTENSIONS

24. Suppose that the apparatus you used in Experiment 6.2, The Decomposition of Water, contained 100 cm³ of water. Using this apparatus, a student collected 57 cm³ of hydrogen and 28 cm³ of oxygen. What fraction of the total volume of water was decomposed?

25. About a gram of salt is placed in a test tube half full of water and shaken; about a gram of citric acid is placed in a test tube filled

with alcohol and shaken; some magnesium carbonate is dropped into half a test tube of dilute sulfuric acid; hydrochloric acid is poured into a dish containing magnesium. In each case the solid disappears, and we say that it "dissolved." However, it is evident that two different kinds of dissolving have occurred. Divide the-experiments into two classes. What did you observe that led you to divide them this way? What do you think you will observe in each case if you evaporate the solution to dryness?

26. A mass of 5.00 g of oxygen combines with 37.2 g of uranium to form uranium oxide.
    a. How many grams of the oxide are formed?
    b. What is the ratio of uranium to oxygen in this compound?
    c. How much oxygen is needed to oxidize completely 100 g of uranium?

27. Discuss the process of boiling an egg in terms of complete and incomplete reactions. What may affect the time required for a complete reaction?

28. What evidence do you have for the following statements?
    a. Zinc chloride is a pure substance and not a mixture.
    b. Water is a pure substance.

29. A student heated a crucible containing 2.15 g of powdered copper until the mass of the contents of the crucible became 2.42 g. The black solid in the crucible was then placed in hydrochloric acid, and 1.08 g of copper remained undissolved in the acid. What is the reacting ratio by mass of copper with oxygen?

30. Hydrochloric acid is a water solution of the compound hydrogen chloride. Suppose there were two bottles containing hydrochloric acid in your lab. How would you determine whether the concentration of hydrogen chloride in each bottle is the same?

31. a. If you heat a piece of blue vitriol (a blue solid), mass decreases and the solid changes to a white powder. Which of the two substances might be an element?
    b. If you dissolve the white powder in water and place an iron nail (iron is an element) in the solution, the nail will become coated with a thin layer of copper. What do you now conclude about the two substances in part (a)? Is either of them an element?

THEMES FOR SHORT ESSAYS

1. "Seeing is believing" is a common saying. But does "not seeing" imply "not believing"? Neither hydrogen nor oxygen is visible in Figure 6.2 or in the laboratory. Is there, perhaps, a form of indirect seeing? Express your thoughts on this subject.

2. Express the main points of Section 6.5 as a debate between Berthollet and Proust.

3. A local newspaper printed a letter from a person who wrote, "Aluminum is plentiful; and if I buy an aluminum can, I can do what I please with it." Write a letter to the editor in reply to this person.

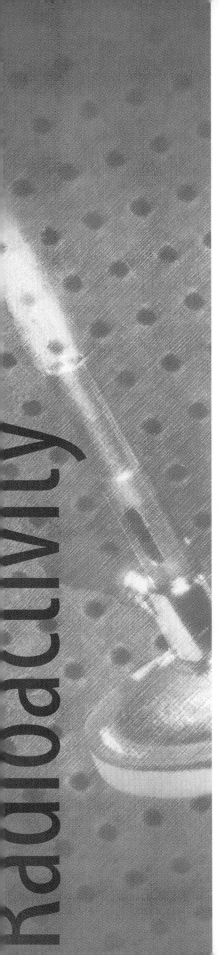

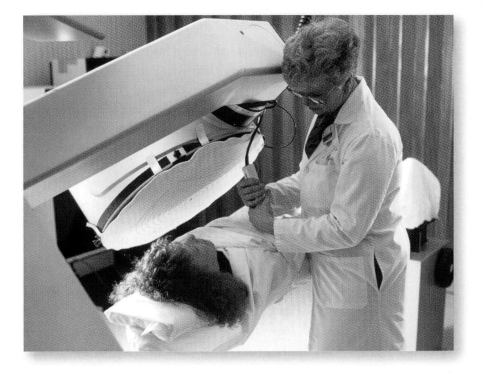

Chapter 7
Radioactivity

7.1 Radioactive Elements

You have now reached a high point in your study of matter. Through simple experiments, you have traced the main steps leading to a recognition of how the material world is put together. The accomplishments of the past few centuries are really impressive: The millions of mixtures around us are now known to be made up of tens of thousands of compounds, which in turn are composed of only about a hundred elements. Compounds have characteristic properties that are different from those of the elements from which they are made. Yet the elements never really disappear. They can always be extracted from their compounds. Elements seem to be permanent.

In science, however, it often happens that a new observation is made just when we think that our understanding of a subject is complete. This observation forces us to admit that our picture is incomplete; some new features must be added.

In the case of elements, some substances that qualified as elements were found to be *not* permanent, but to change on their own into other elements. The ground-breaking experiment leading to this discovery was made by the French physicist Henri Becquerel about a hundred years ago. We shall now describe a modern version of his experiment.

Six plastic boxes containing samples of different materials were placed on photographic film. The film was wrapped in black paper to shield it from light. After three days the film was developed. Figure 7.1 shows where the boxes were placed. Figure 7.2 shows three white squares on the developed film where boxes *A*, *C*, and *E* were placed. Apparently, the

Figure 7.1
Six different substances in small plastic boxes are placed on photographic film enclosed in black paper. The boxes are left in position for three days before the film is developed and the photograph produced.

Figure 7.2
The finished photograph from Figure 7.1. The three white squares appear where boxes *A*, *C*, and *E* were placed.

material in these boxes gave off something that was able to get through the black paper and affect the film in the way that light does.

The elements in the boxes (all in compounds, except in box *F*) are listed in Table 7.1. Boxes A and B had two elements in common: sulfur and oxygen. However, box *A*, not box *B*, affected the film. This suggests that uranium was the element that emitted, or *radiated*, something. This conclusion is supported by the fact that box *C*, which also contained uranium, left a mark on the photographic film.

Similarly, a comparison of the contents of boxes *D* and *E* suggests that the thorium in box *E* is the cause of the mark on the film.

Table 7.1

Box	Elements
A	Uranium, sulfur, oxygen
B	Sodium, sulfur, oxygen
C	Uranium, nitrogen, oxygen
D	Sodium, nitrogen, oxygen
E	Thorium, nitrogen, oxygen
F	Sulfur

Photographic film is not the only tool that distinguishes boxes *A*, *C*, and *E* from the others. If we place any one of these boxes near an instrument called a Geiger counter (Figure 7.3), the counter will count or produce clicks that can be counted. Boxes *B*, *D*, and *F* do not affect the counter at all.

There are other elements listed in Table 6.2 that radiate something that affects a photographic plate and a Geiger counter. These elements are called *radioactive* elements.

†1. Elements *X*, *Y*, and *Z* form compounds *XY*, *XZ*, and *YZ*. Compounds *XY* and *YZ* are radioactive but compound *XZ* is not. Which element is radioactive?

2. A piece of magnesium placed in hydrochloric acid causes hydrogen to be released. Evaporation of the resulting solution leaves behind a white solid. A similar reaction occurs when a piece of uranium is placed in hydrochloric acid. Do you expect this white solid to be radioactive?

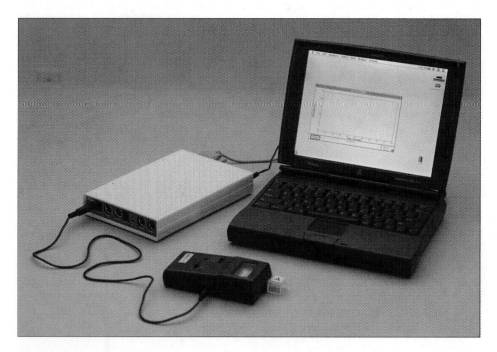

Figure 7.3
A Geiger counter used to detect radiation from radioactive substances.
The detector, next to the yellow sample, is connected to a computer,
which serves as a recorder.

7.2 Radioactive Decomposition

What happens to a radioactive element when it radiates? This is a hard question to answer. Some elements, such as uranium, radiate so weakly that it is extremely difficult to discover what is happening to them. Other radioactive elements, such as polonium, emit such intense radiation that it is a simple task to determine what happens. However, because of its intense radiation, even a milligram (10^{-3} g) of polonium is a health hazard unless special precautions are taken.

Despite the difficulties, both weakly and strongly radiating elements have been studied using a variety of methods. One method, spectral analysis, will be discussed in Chapter 8. In the case of polonium, spectral analysis shows that the radiation is associated with changing polonium into lead and helium. Spectral analysis shows that other radioactive elements also change into different elements. In many cases, helium is one of the elements that is produced. Whether helium is produced or not, we refer to this process as *radioactive decomposition* or *radioactive decay*.

Is radioactive decomposition different from the decomposition of water or other compounds? You saw that the products of the decomposition of water could be recombined to form the original compound. Lime, which Lavoisier thought to be an element, is very difficult to decompose into calcium and oxygen even by electrolysis. Nevertheless, calcium and oxygen easily recombine to form lime. It has been possible to recombine all the elements that have been obtained from compounds. However, no one has been able to recombine the products of radioactive decomposition by using the methods that are effective in recombining elements into compounds.

What happens to a radioactive element when it is heated? Does heat change the rate at which it emits radiation? In many of the experiments you have done and read about in this course, temperature has had a big effect on what happened and how fast it happened. Baking soda emitted a gas only when you heated it. Hydrogen does not burn by itself in air, but when it is heated with a burning match, it catches fire very easily—sometimes explosively.

The effect of temperature on the rate at which radiation is emitted by radioactive substances has been studied. Samples of the substances were heated to very high temperatures and cooled to very low temperatures. Temperature had no effect on the radiation from a radioactive substance.

To sum up, radioactive elements have all the properties of elements that you studied in the preceding chapters. They form compounds with constant proportions, and they have characteristic densities, melting points, and boiling points. They cannot be decomposed by ordinary heat, electricity, reaction with acids, and the like. They differ from non-radioactive elements in that they affect a photographic plate and decompose into other elements. The rate at which they decompose cannot be changed by any of the means that affect the rate at which compounds decompose. That is why these substances are called elements; but to set them apart, we call them radioactive elements.

3. **What are the two most important differences between the following two reactions?**
   a. **Water → hydrogen + oxygen**
   b. **Polonium → lead + helium**

†4. **A radioactive sample at 20°C is placed near a Geiger counter. The counter records 1.0×10^2 counts per minute. The temperature of the sample is then raised to 100°C. What does the counter then record?**

EXPERIMENT
7.3 Radioactive Background

When a Geiger counter is placed near a radioactive substance it registers decays. What do you expect a counter to read if no known radioactive substance is nearby?

When there is only one Geiger counter per class, each of you may be assigned different tasks in this class experiment. Set up a Geiger counter in your classroom and let it count for 1.0 minute. Reset the counter and take two or three additional readings for one minute each.

• What was the number of counts in each reading? Were the readings all the same?

Let the counter run for 30 minutes.

• What was the average number of counts per minute over the 30-minute interval?

The average number of counts per minute registered by a counter when no obvious radioactive source is present is called the *radioactive background*, or *background* for short.

The background may depend on location and, in a closed space, on ventilation. Repeat the background measurement in the basement or other room chosen by your teacher.

• What was the background in the basement or other room?

5. A student measured the radioactive background in two locations. At location A she observed 50 counts in 2.0 minutes, and at location B, 150 counts in 10.0 minutes. Which location had the higher background?

6. Six readings of a Geiger counter were taken over 1-minute intervals. The numbers of counts per minute were: 12, 14, 15, 9, 7, and 18. What do you think the result of the next reading would have been? Why?

EXPERIMENT
7.4 Collecting Radioactive Material on a Filter

You may or may not be able to tell whether the background in the basement is significantly higher than in your classroom. But you can find out if the background radiation in the basement has the same origin as the background radiation upstairs. The radiation can come down from the sky, it can come up from inside the earth, or it can come from radioactive material floating in the air.

Suppose the source is dust in the air. It may be possible to catch some of this dust by passing a large volume of air through a filter. You can do this by fastening a piece of paper towel around the end of the tube of a vacuum cleaner (Figure 7.4). After the air has flowed through the filter for a time set by your teacher, turn off the vacuum cleaner and place the

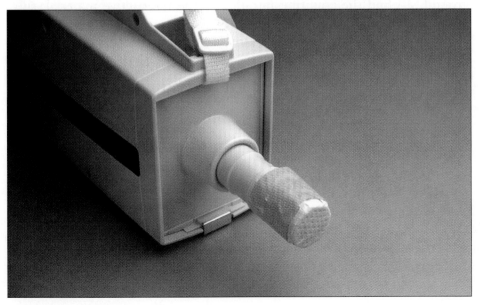

Figure 7.4
The end of a tube of a vacuum cleaner covered with a piece of paper towel serving as a filter.

Figure 7.5
The loaded filter and
the Geiger counter.

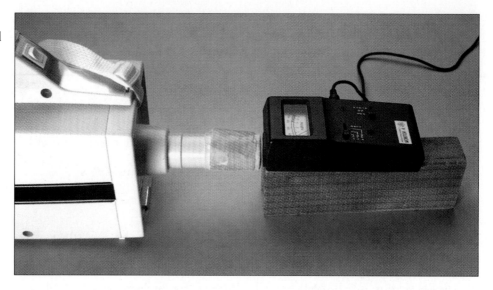

Geiger counter in front of the filter for 3 or 4 minutes (Figure 7.5). Dividing the number of counts by the time during which they were collected gives the *counting rate*.

- How does the counting rate compare with the background when classroom air passed through the filter?
- How does the counting rate compare with the background in the basement when basement air passed through the filter?
- What do you conclude about the source of the extra counts?

The counting rate that you have measured so far includes the background. To find the counting rate due to the radioactive material, subtract the background from the measured counting rate. The difference is called the *net counting rate*. Whenever the measured counting rate is large compared to the background, the net counting rate will be close to the measured one. However, when the measured counting rate is close to the background, the net counting rate will be close to zero.

7. A class had the following results when collecting radioactive material on a filter: in a closed basement closet, 412 counts in 2.00 minutes; on an open terrace, 36 counts in 2.00 minutes. A 19 counts/min background was found from an earlier run that lasted 30.0 minutes.

   a. What were the two measured counting rates in counts per minute?

   b. What were the two net counting rates?

   c. If one of the net counting rates were negative, how would you explain such a result?

EXPERIMENT
7.5 Absorption and Decay

Keeping the filter and the counter in fixed positions, hold one, two, or more pieces of paper between them, and measure the counting rate. For reasons that will become clear very shortly, these measurements are best done in rapid succession.

- Does the paper affect the net counting rate? If so, how?
- Which has the greater effect on the net counting rate, a few pieces of paper or a sheet of aluminum?

Measure the counting rate two more times, at about 20 minutes and 40 minutes after the first measurement. (The exact times are not important as long as you keep a record of what they were.) Taking the time of the first measurement as zero, plot the net counting rate as a function of time. Your data may look somewhat like those in Figure 7.6: the counting rate decreases with time, but apparently in an irregular way. Can we represent such data by a smooth curve?

There is a difference between plotting radioactivity measurements and, say, mass measurements. Massing a test tube several times will give the same results within the sensitivity of the balance. For example, the results may vary between 23.24 g and 23.26 g. However, a mass of

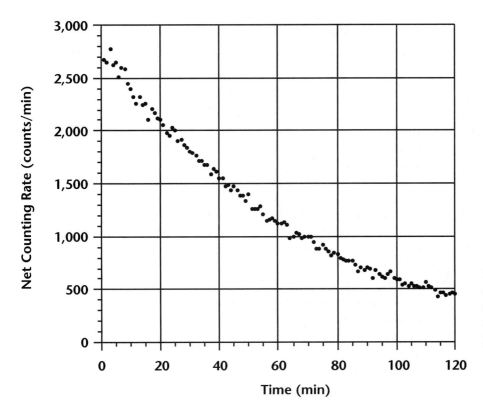

Figure 7.6
Actual class data for the net counting rate for radioactive material collected on a paper filter.

Figure 7.7
A smooth line drawn
close to the data points
in Figure 7.6. The fact
that the first and last
points are closest to the
curve is accidental.

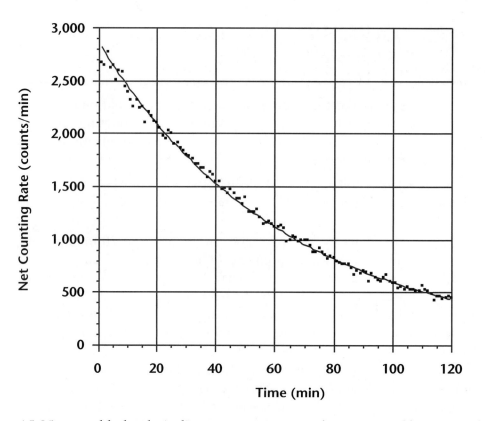

15.85 g would clearly indicate an error in reading or a malfunction of
the balance. The results of counting-rate measurements are quite different. Background measurements over short time intervals vary considerably. The same is true for the net decay rate of a radioactive
substance. Another run under the same conditions would produce different numbers showing only the same general trend.

Therefore, it makes no sense to try to draw a curve through all the
data points. A smooth curve that has about the same number of data
points on each side is likely to fit different sets of data and so be more
representative. Figure 7.7 shows such a curve. Other curves close to the
one in Figure 7.7 would also be acceptable.

Draw a smooth curve for your net counting rate as a function of time.

• How long did it take for the net counting rate to decrease to half its
original value? From half the original value to a quarter?

• Do you see any relation between these two time intervals?

• What net counting rate do you expect to find after one day?

8. In Figure 7.7, how long did it take for the net counting rate to
decrease to half its original value? From half the original value to a
quarter?

9. Figure A is a graph of the net counting rate as a function of time for a sample of polonium.
   a. How long did it take for the counting rate to decrease to half its original value? From half the original value to a quarter?
   b. How are these two time intervals related?

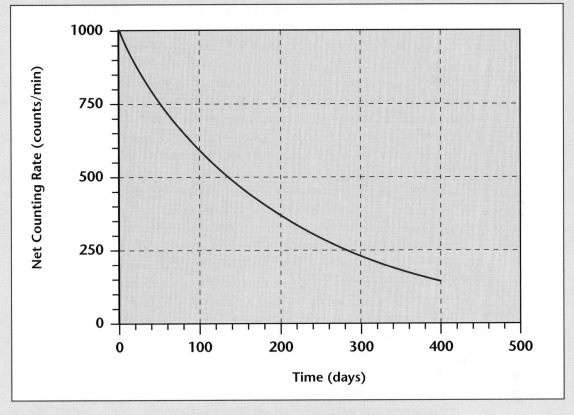

10. Suppose a filter used to collect radioactive dust is made of two layers of paper towel. The two pieces of paper are placed in front of identical Geiger counters for the same length of time. Which layer, the outer or the inner, will produce the greater number of counts? Explain your reasoning.

7.6 Radioactivity and Health

When the radiation from radioactive substances hits photographic film, it makes some changes in the film. If it did not, the film could not serve as a detector. When the same radiation hits other materials, it also causes some changes. If the material hit is a human body, the changes can be harmful if the radiation is sufficiently strong. How deep into the body does the radiation found in basements penetrate?

The thickness of a layer that is needed to absorb most of the radiation from a particular source depends on the source and the absorber. Some of the radiation found in basements cannot even enter a common Geiger counter, although the front of the counter has a very thin cover. This radiation can be detected only by special counters. Some radioactive materials must be placed behind a thick wall to provide effective shielding.

What is the danger from the radiation found in basements? The observation that even a piece of paper absorbs a noticeable fraction of the radiation suggests that much will be stopped by the clothing and the skin. However, our lungs have no skin to protect them. They have a large surface area for the exchange of oxygen and carbon dioxide. Between inhaling and exhaling, radioactive material in the air can be deposited on the surface of the lungs. If the concentration of radioactive material in the air is large enough and the air is breathed over a long enough time, the danger of lung cancer cannot be ignored.

Why is the danger mostly in basements? A radioactive gas called *radon* originates inside the earth. Radon is an element, which, like helium and neon, does not form compounds. If it reaches the surface of the earth, it will enter the atmosphere. If it reaches a basement, its concentration will rise, because basements are entered less often and are not as well ventilated as upstairs rooms. The radiation emitted by radon is of the kind that is not detected by most Geiger counters. What you observed was the radiation emitted by the decay products of radon.

In general, the concentration of radioactive material in the air of a basement is small. This is why we needed a vacuum cleaner to collect enough radioactive material from a large volume of air.

Just as radiation from radioactive substances can be harmful to humans, it also destroys the bacteria that cause fruits to rot and milk to spoil. In recent years irradiation has increasingly been used to preserve food products. Products so preserved contain no chemical additives (Figure 7.8).

Figure 7.8
A comparison between irradiated (right) and nonirradiated strawberries (left). Both samples were kept under refrigeration for two weeks beneath a plastic cover. (*Courtesy International Atomic Energy Agency*)

Figure 7.9

A photographic plate exposed to a tiny amount of polonium.

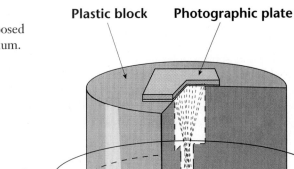

Plastic block   Photographic plate

Source

7.7 A Closer Look at Radioactivity

So far we have observed some general features of radioactive decay. Now we shall take a closer look at how it happens. We begin with the effect of a radioactive element on photographic film. Look again at Figure 7.2. The two white squares and the gray square appear uniform.

A similar experiment was done using a very small sample of polonium. Three photographic plates were exposed to radiation in the arrangement shown in Figure 7.9. The plates were exposed for 66, 94, and 144 hours, respectively. When the developed plates were examined, the first one showed a very light uniform gray spot, the second had a darker shade, and the third was still darker.

When the center of each of the spots was photographed through a high-powered microscope with a magnification of about 10^3, none of them appeared to be a uniform shade (Figure 7.10). In all of the photographs, the dots are about equal in size and blackness, and form no particular pattern. They appear to be distributed in a random way. On the more exposed plates, there are, on average, more black dots in each square.

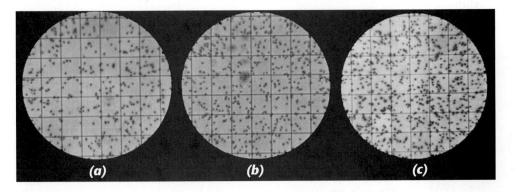

(a)   (b)   (c)

Figure 7.10

Three photographic negatives obtained with the apparatus shown in Figure 7.9. The three pictures show very small, equal-sized areas of the negatives, magnified a thousand times. Negatives (a), (b), and (c) were exposed for 66, 94, and 144 hours, respectively.

Figure 7.11

A simple cloud chamber. The chamber itself is a cylindrical plastic box resting on a block of Dry Ice. A dark felt band soaked in isopropanol encircles the inside of the top of the chamber. The bottom of the chamber is painted black so that the white fog track will be clearly visible from above. The chamber is illuminated from the side. The radioactive source is on the tip of the needle.

When a Geiger counter is close to a radioactive sample, you hear distinct clicks. The clicks occur at random. Even after listening for hours, you would not be able to predict when the next click would come. We can be confident about only the average number of clicks in any given time interval.

A more vivid picture of radioactive decay is provided by a device called a *cloud chamber* (Figure 7.11). Figure 7.12 shows a photograph of a cloud chamber that contains a sample of polonium. You can see individual tracks originating from the sample. The tracks are formed instantaneously—as far as one can tell just by looking. Nevertheless, it seems reasonable that they are formed by very fast particles flying off from the radioactive source.

Watching a cloud chamber in operation, you will notice that the tracks shoot off irregularly. Just as with the black dots on the film and the clicks of the counter, you will not be able to predict when and in which direction the next track will appear. Only the average number of tracks produced in a long time interval can be predicted from past observations.

Figure 7.12

Fog tracks produced in a cloud chamber by polonium. The preparation of the cloud chamber and the formation of the tracks are shown in the *IPS* video "Radioactive Substances II."

This kind of behavior is characteristic of events that happen by chance. For example, suppose you throw 10 dice many times. In each throw of 10 dice, you are likely to find one or two sixes. In some cases you will find three or more sixes, or none at all. You have no way of telling what the next throw will produce. However, the average number of sixes in many throws of ten dice can be safely predicted to be 10/6.

The manner in which film, Geiger counters, and especially cloud chambers record radioactive decomposition suggests that this process takes place in discrete, or countable, steps. Each black dot, click, or track signals the arrival of a small particle emitted by the radioactive source. Combining this idea with the fact that elements are produced during radioactive decomposition leads to a far-reaching conclusion: Elements are made up of discrete units, or some sort of tiny particles, called *atoms*.

Historically, the idea that elements are made up of atoms preceded the discovery of radioactivity. We have chosen to introduce this idea through radioactivity because radioactivity enables us to count individual events in which elements are changed.

†11. Can you tell whether the tracks in Figure 7.12 are slanting upward, are slanting downward, or are horizontal? What assumption are you making in arriving at your answer?

12. Summarize the similarities between radioactive decay as observed with a cloud chamber and with a Geiger counter.

13. State in your own words why radioactive decomposition suggests that elements are made of tiny particles.

FOR REVIEW, APPLICATIONS, AND EXTENSIONS

14. A plant absorbs various substances through its roots. Different elements of these substances concentrate in different parts of the plant. Suppose one of these elements is radioactive. How would you determine in what parts of the plant it concentrates?

15. a. With a paper towel in place, it took a vacuum cleaner 32 seconds to fill a 160-liter plastic bag with air. How much air passed through the filter in 1 second? In 1 minute?
    b. Suppose that you used the same kind of vacuum cleaner in Section 7.4. How much air would pass through the filter in the time you collected the radioactive material?

16. a. Breathe normally into a small plastic bag and estimate the volume of air that you exhale in one breath.

    b. Count how many times you exhale per minute (breathing normally).

    c. What volume of air passes through your lungs each minute?

    d. Suppose you stayed in the basement for the same period of time that it took you to collect the radioactive material in Section 7.4. How much air would pass through your lungs?

    e. Suppose that your lungs act like the combination of filter and counter used in Section 7.4. What would be the "counting rate" of your lungs due to breathing the basement air?

17. In the *IPS* video "Radioactive Substances I," the Geiger counter counted much faster when box *A* was placed next to it than when box *E* was. Is this what you would have expected on the basis of the brightness of the white squares in Figure 7.2? Why or why not?

18. The *IPS* video "Radioactive Substances II" shows tracks being produced in a cloud chamber. From what you see in the loop, can you be sure that the tracks start at the source rather than end there? What does this tell you about the speed of the particles that leave the tracks?

THEMES FOR SHORT ESSAYS

1. Suppose a law were passed making it illegal to sell a house in which radon is present in the basement. Write your thoughts about this proposed law.

2. The use of fluoride to reduce tooth decay was quite controversial in the past. Today, irradiating fruits and vegetables to reduce spoilage is controversial. Interview your teachers, parents, and friends to assess their feelings on this issue. Try to determine the basis for these feelings.

The Atomic Model of Matter

8.1 A Model

Suppose someone hands you a sealed can. You shake it, and hear and feel something sloshing around inside. From this simple experiment you can form a mental picture—a *model*—of what is inside the can. You conclude that the can contains a liquid. You have no knowledge about the color, taste, or odor of the liquid. But you are sure that the contents of the can has a characteristic property of liquids—it sloshes around when you shake the can. The "liquid model" of the can accounts for your observation.

However, for a model to be useful, it has to offer more than just a convenient way of accounting for known facts. A model must also enable us to make testable predictions. For example, from the "liquid model" of the can, we can make the following prediction: If you punch a small hole in the bottom of the can, liquid will drip out.

The simple model we have made for the behavior of the can was the result of just one experiment, shaking the can. It led to only one rather obvious prediction. Before we go on to develop an atomic model of matter, you will do an experiment with a "black box." This experiment will give you an opportunity to develop a richer model than the one offered by the can. It will also make it easier for you to understand the atomic model of matter.

†1. A good model enables us to summarize and account for the facts we learned from observation and experiment. What else should a good model do for us?

2. Suppose you have two eggs in the refrigerator. One is raw and the other is hard-boiled, but you do not know which is which. You spin each egg. One egg spins freely. The other egg comes to rest after very few rotations.
   a. Suggest a model that will account for the different behavior of the two eggs.
   b. On the basis of your model, what do you predict about the spinning of the two eggs after you boil both for 10 minutes?
   c. Check your prediction at home.

 EXPERIMENT
8.2 A Black Box

All the boxes you and your classmates will use in this experiment are the same (Figure 8.1). The first step is to find out as much as you can about these boxes without pulling out the rods, and, of course, without

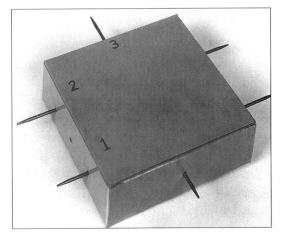

Figure 8.1
A "black box."

opening the boxes. Look at one of them, shake it lightly, tilt it back and forth in various directions, and listen to the sounds. Writing down your observations will help you compare notes with your classmates. You will then be ready to arrive jointly at a model, make predictions, and test them.

Do all the experiments you can think of, short of pulling out the rods or opening the box. Try to imagine in a general way what is inside the box that could account for your observations. This will be your model for the box. Do not be distracted by details. Do not, for example, try to name the objects inside the box; only describe them by the properties that you have found in your experiments. If you hear something sliding on one of the rods, it might be a washer or a ring. The important point is that it is something with a hole in it through which the rod passes.

After you and your classmates have made models that account for your observations, predict what will happen when you pull out a particular rod. Also predict how this will affect the results of the tests you performed earlier. (Be sure to write down your predictions so that you can check them.)

Now you or one of your classmates may remove this rod from one of the boxes. Leave all the rods in place in the other boxes. If what happens confirms your prediction, you can use one of the other boxes to test predictions about what would happen if another rod were pulled out first. If, however, your first prediction was not confirmed, modify your model before experimenting further. Continue this process until you have arrived at a model in which you have confidence.

3. **a. In investigating the black box, you did certain things to it that you could undo. Give some examples.**

    b. What did you do to the black box that you could not undo?

    c. Does dissolving zinc in hydrochloric acid resemble any of the kinds of tests you mentioned in part (a) or part (b)?

8.3 The Atomic Model of Matter

Let us now begin to build an atomic model for matter by reviewing some of the common properties of elements, mixtures, and compounds. The assumptions we introduce into the model must account for these properties. Except in the last two sections of this chapter, we shall concentrate on stable, nonradioactive matter.

First, we consider elements. Different samples of an element have the same characteristic properties. To account for this, we shall assume that an element is made up of tiny particles of only one kind. We call these particles *atoms*. Different elements are made up of different kinds of atoms. Atoms cannot be seen even with strong microscopes. So they must be very small and there must be very many of them in any sample large enough for us to examine.

At this point we do not know how atoms of different elements differ from each other. But we can use familiar objects to illustrate the idea. Figure 8.2 shows a group of paper fasteners and a group of rubber rings. Think of a fastener as an atom of element Fs, and think of a ring as an atom of element R. Of course, nobody believes that an atom looks like a fastener or a ring. But this simple analogy illustrates the basic idea of the atomic model: All atoms of one element are alike, but they are different from the atoms of all other elements.

Second, we consider mixtures. Mixtures of elements can be made in varying proportions. Furthermore, the characteristic properties of a mixture can be made to vary widely by varying the amounts of the elements being mixed. And at least some of the characteristic properties of the individual elements (for example, color and the ability to react with other substances) are usually present in a mixture.

We can account for the behavior of mixtures by assuming that in mixtures the atoms of the individual elements remain unchanged. The numbers and masses of the atoms remain the same. The atoms just get mixed like marbles of different colors. Note that with this assumption the model guarantees that mass is conserved when mixtures are made and separated. Figure 8.3 provides a visualization of a mixture of two elements in terms of fasteners and rings.

Third, we consider compounds. When elements form compounds, they combine only in a definite ratio. You saw evidence for the law of constant proportions in the burning of hydrogen with oxygen and in the synthesis of zinc chloride. Furthermore, the compound produced in each case had its own set of characteristic properties, usually quite different from those of the reacting elements. For example, the characteristic properties of carbon dioxide are entirely different from those of the carbon and oxygen that combine to make it.

Figure 8.2
Paper fasteners representing atoms of "element" Fs, and rubber rings
representing atoms of "element" R.

To account for the law of constant proportions a new assumption
must be added to the atomic model. We assume that when a compound
is formed, each atom of one element attaches to a fixed number of atoms
of the other elements in a particular pattern. A simple illustration of

Figure 8.3
Two mixtures of the "elements" Fs and R in very different proportions.

Figure 8.4
The compound FsR made by attaching one "atom" of Fs to one "atom"
of R. Note the excess of Fs.

compound formation is provided by the attachment of one "atom" of Fs to one "atom" of R (Figure 8.4). Note that the excess of Fs did not "react." Thus the idea of attachment leads directly to the law of constant proportions.

4. Suppose that *M* atoms of mercury combine with *N* atoms of oxygen to form mercury oxide.
   a. What total number of atoms would you expect there to be in the mercury oxide produced?
   b. If the mercury oxide produced is then decomposed by heating to form gaseous mercury and oxygen, how many atoms of mercury and how many atoms of oxygen would you expect to find?
   c. Would your answer to (b) be different if you had condensed the gases to liquid mercury and liquid oxygen?

"EXPERIMENT"
8.4 Constant Composition Using Fasteners and Rings

To illustrate what the atomic model says about the formation of compounds, you will use the "elements" Fs and R introduced in the preceding section. When you fit rings on fasteners as shown in Figure 8.4, the number of rings and the number of fasteners do not change. Hence the total mass does not change. The atomic model accounts for the conservation of mass.

Making a compound from these objects is similar to forming copper oxide from copper and oxygen (Experiment 6.6, A Reaction with Copper). In that experiment you worked with large numbers of invisible atoms. In this experiment you will illustrate the process by using many visible "atoms" to form a sample of a compound.

• What is the mass of all the Fs that you have?

Make as much of the "compound" FsR as your supply of "atoms" allows by putting one ring on each fastener (Figure 8.4). When we write "FsR", we mean a compound made up of one "atom" of Fs for each "atom" of R.

• What is the mass of the product you have synthesized?

If you have an excess of Fs, find its mass.

• How much Fs reacted with R?

• What is the ratio of the mass of R to the mass of product in your sample of the "compound?"

Compare your results with those of your classmates.

• Does the ratio of the mass of R to the mass of product depend on how big a sample you make?

• Does the model as illustrated by fasteners and rings agree with the law of constant proportions?

• Would the ratio have been the same if you had used heavier rings?

Suppose you "decomposed" your "compound" into pure Fs and R by taking the rings off the fasteners. You would get back all the "atoms" you started with.

• If you then measured the masses of the "elements" Fs and R, would the model agree with the law of conservation of mass?

†5. **Does the experiment with rubber rings and paper fasteners give you any information about the shape of an atom?**

6. **In the synthesis of water described in Section 6.3, different volumes of hydrogen were mixed with a constant volume of oxygen. Then the mixture was ignited by a spark. In the first case, there was some oxygen left over; in the second case, nearly all the oxygen and the hydrogen reacted; and in the third case, hydrogen was left over. Describe these results in terms of the atomic model of matter.**

7. Suppose that element A can form a compound with element B but not with element C. Element B can form a compound with C. How could you use fasteners, rings, and washers to represent these elements?

8. Describe the synthesis of zinc chloride (Experiment 6.4, The Synthesis of Zinc Chloride) in terms of fasteners, rings, and washers. (Hint: Hydrochloric acid contains hydrogen and chlorine.)

8.5 Molecules

When two elements form a compound, their atoms attach themselves in a given ratio. This is as far as we have come in terms of the atomic model. Let us now try to be more specific.

Figure 8.5 shows several possible arrangements for a compound containing equal numbers of atoms of two elements. In Figure 8.5(a) the atoms of the two elements are clearly paired off, with space between each pair. In Figure 8.5(b) the atoms form clusters of four. In Figure 8.5(c) and (d) we can no longer distinguish groups or clusters that contain

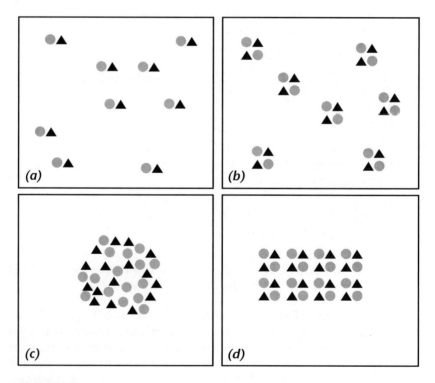

Figure 8.5
Possible attachments of the atoms of two elements in a compound that contains equal numbers of each kind of atom.

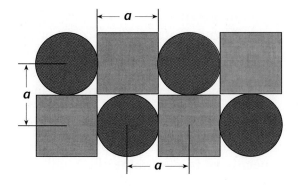

Figure 8.6
In a solid, the distance between the centers of atoms equals their size.

a fixed number of atoms. In Figure 8.5(c) there is no particular order to the atoms, whereas in Figure 8.5(d) they are arranged in neat rows. Which of these pictures fits the arrangement of atoms in a real compound? To answer this question, we go back to what we already know about the behavior of matter in bulk.

Anybody who has ever used a bicycle pump knows that it is quite easy to compress air, that is, to reduce its volume. The same is true for all gases. But try to reduce the volume of a small crystal of sodium chloride even by 1 percent. Without special equipment you will not succeed. A cube of copper will behave similarly.

While gases are quite compressible, solids are almost incompressible. We can account for this by assuming that atoms behave like little hard objects whose size remains fixed. In a solid, the atoms are close together; they are all "touching" one another. This means that the distance between the centers of the atoms equals the size of the atoms themselves (Figure 8.6). Such a picture suggests that solids are hard to compress. In a gas, on the other hand, the atoms do not touch each other. They are far apart relative to their size; they can easily be pushed closer together to occupy a smaller volume (Figure 8.7).

Figure 8.7
The large box represents a volume of helium gas at atmospheric pressure and room temperature. The spheres in the box represent helium atoms. The small box shows the volume occupied by the same helium atoms when helium is liquified.

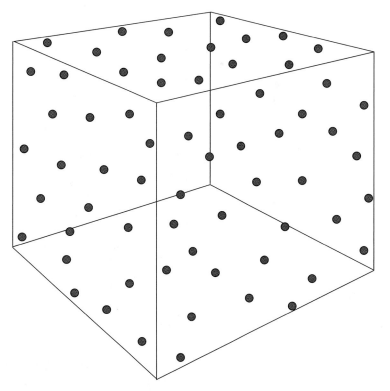

A liquid or solid compound AB can be represented by Figure 8.5(c) or (d). A gas AB can be represented by Figure 8.5(a) or (b). Groups or clusters containing a fixed number of atoms are called *molecules*. Each molecule of a given compound contains the same number of atoms. (A substance with more than one kind of molecule is not a pure substance.)

Compounds are not the only substances made of molecules. The atoms of an element can also form clusters or molecules, with only one kind of atom in each cluster. The analysis of many experiments and a long chain of arguments are necessary to determine the numbers of atoms in the molecules of different gases. In this course we do not present them. We shall simply state the results for some of the substances we have used as illustrations.

In gaseous hydrogen, the atoms cluster in pairs. This is the reason why a molecule of hydrogen gas is written as H_2. The subscript indicates the number of atoms of a given element. The atoms of nitrogen and oxygen also cluster in pairs. A molecule of carbon dioxide (CO_2) is a cluster of two oxygen atoms and one carbon atom. The atoms of helium gas (He) and radon gas (Rn) do not form clusters. (Elements like these are said to have *monatomic* molecules.)

Every sample of a given compound contains the same elements in the same ratio by mass. However, two substances that have the same elements in the same ratio by mass need not be the same compound. For example, the two oxides of nitrogen, NO_2 and N_2O_4, have the same ratio of nitrogen and oxygen. However, the molecules of NO_2 have three atoms, and those of N_2O_4 have six atoms.

Another example is provided by ethyne (also called acetylene) and benzene, two compounds of carbon and hydrogen with the same mass ratio. A molecule of ethyne, C_2H_2, has four atoms. A molecule of benzene, C_6H_6, has twelve atoms. The two compounds have very different properties.

When atoms are arranged as in Figure 8.5(d), it is meaningless to talk about clusters or molecules. Describing such a solid as a compound AB tells us only that it contains equal numbers of the two kinds of atoms. Thus, the formula NaCl, when applied to solid sodium chloride, means only that the crystal contains equal numbers of sodium atoms and chlorine atoms. Similarly, the formula H_2O applied to water states only the ratio of hydrogen and oxygen atoms in the liquid. However, a molecule of water vapor is accurately described by H_2O.

9. **Suppose that you find two solid compounds, one more compressible than the other. Which compound do you think is better described by Figure 8.5(c) and which by Figure 8.5(d)?**

EXPERIMENT
8.6 Flame Tests of Some Elements

As you saw in Chapter 6, some compounds are hard to break up into their elements. However, by holding small samples in a flame, even such compounds can be made to reveal the elements they contain.

Place small quantities of different compounds, each of which contains sodium, on tiny loops of nichrome wire and hold them in a flame. Record the color you see for each of the compounds.

Next, try the same experiment using copper and compounds containing copper. Repeat the same procedure with compounds containing strontium, lithium, and calcium.

- How can you recognize sodium in a compound by doing such a flame test?

- How can you recognize copper?

- Can you distinguish among strontium, lithium, and calcium by a flame test?

Although the characteristic properties of compounds are different from those of the elements that make them up, the color of the flame is due to the elements. We can understand the appearance of the colors in terms of the atomic model in the following way: The flame is not hot enough to break up the entire sample. But it can shake off some atoms from their compounds. Once these atoms are separated, they emit light of a color specific to them. It does not matter anymore from what compound they were released.

> 10. When you hold a small amount of sodium chloride in a flame, you observe that the flame is bright yellow. What could you do to be sure that the color is due to sodium and not to chlorine?

> 11. When you spill a few drops of soup or milk on a pale-blue gas flame when cooking, the flame changes to a mixture of colors. Yellow is the most intense color. What do you conclude from this observation?

EXPERIMENT
8.7 Spectra of Some Elements

As you probably just found out, some elements are easy to identify by a flame test. Many other elements are not so easily identified. For such elements we need to separate the mixed colors in order to detect slight differences. The first step is to spread out the various colors in the light in much the same way they appear in a rainbow. This spread of colors is called a *spectrum* (plural: *spectra*).

You can produce a spectrum using a simple grating spectroscope. This devise consists of a tube with a slit at one end and a transparent plastic disk at the other. The disk has many parallel lines on it.

Hold the end with the plastic grating next to your eye and look at several light sources provided by you teacher. When the light source is a thin tube, you may find it more convenient to look at the light with the slit removed.

- What is the major difference between the spectrum of an ordinary light bulb and the spectrum of any of the gas tubes?

- What are the differences between the spectra of the different gas tubes?

8.8 Spectral Analysis

We mentioned compounds of calcium, lithium, and strontium without specifying which compound we were talking about. This may give you the impression that the spectrum of only one of the elements in a compound can be observed. This is not so. The flame of your burner is hot enough to produce the spectra of sodium, lithium, calcium, copper, and a few other elements. It is not hot enough to produce the spectra of elements such as

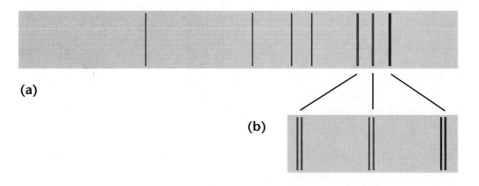

(a)

(b)

Figure 8.8

(a) The spectrum of sodium. The spectrum was in black and white; the color was added later. Therefore, the positions of the lines are accurate, but their colors are not. The line in the yellow is extremely intense. It has been photographed through an absorbing filter so it could be seen at the same time as the other lines. The close pair of almost invisible violet lines arises from a potassium impurity. (b) A part of the sodium spectrum has been photographed with a spectroscope that spreads out the light more, enabling us to see more detail. It shows that the yellow line is really made up of two lines that are very close together.

oxygen and chlorine. However, the very high temperatures generated by an electric arc will produce the spectra of all the elements in a compound.

Under such conditions the resulting spectrum is no longer simple. It will most likely contain complicated patterns of many closely spaced lines. Yet each element gives out its own spectrum, which is different from that of any other element. It takes accurate measurements of the positions of spectral lines to identify an element. Once this has been done, the presence of any element has been definitely established.

With a good instrument, the sodium spectrum looks like Figure 8.8. This shows that the yellow of the sodium flame is not just any yellow. It is a very specific color indeed, which has its own place in the spectrum. No other element has a line at this place in the spectrum. The presence of this particular pair of lines always means that sodium is present in the light source. Even if the yellow color in the flame is masked by many other colors, the spectroscope will reveal the presence of sodium.

Figure 8.9 illustrates spectra obtained from compounds of calcium, lithium, and strontium. All these elements give flame tests of nearly the same color. However, each has its own set of characteristic spectral lines when viewed through a spectroscope.

Unlike the measurement of melting point or density, spectral analysis or spectroscopy, can be done on very distant objects like our sun and other stars. Analysis of sunlight was one of the very early uses of spectroscopy. Most of the spectral lines observed in sunlight could also be produced with known materials in the laboratory. However, during a solar eclipse in 1868, a new set of spectral lines was found in the spectrum of the light

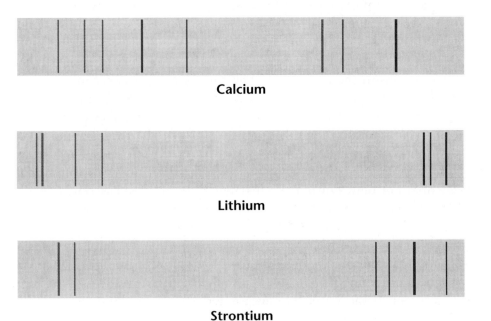

Calcium

Lithium

Figure 8.9
The brightest lines in the spectrum of calcium, lithium, and strontium.

Strontium

Figure 8.10

A comparison between the spectrum of pure helium (upper half) and the spectrum produced in an evacuated tube containing polonium (lower half). The two faint lines close together in the lower half are polonium lines.

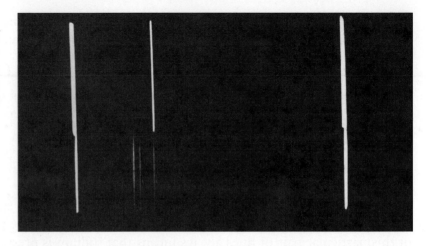

coming from the edge of the sun. This set of lines had never been seen before, and could not be produced in the laboratory with any element known at the time. The lines were therefore thought to be from a new element, which was named "helium" after the Greek word for sun. Eventually, helium was also detected on Earth through spectral analysis.

In Section 7.2 we mentioned that the elements produced in the radioactive decomposition of polonium were identified by spectral analysis. The formation of helium in this process was demonstrated for use in this course by the following procedure:

The upper half of Figure 8.10 shows the spectrum of helium contained in a tube like one you may have seen in the classroom. The lower half shows the spectrum obtained from a tube that was prepared this way: 4.5×10^{-3} g of polonium was deposited inside, the air was pumped out, and the tube was sealed. Thirteen days later the spectrum was photographed.

Note that the three lines in the upper half of Figure 8.10 appear at the same positions in the lower half. The exposure was weaker in the lower half (the lines are thinner); this is why the line near the middle is barely visible in the lower half. The mass of the helium in this experiment is calculated in Section 9.6 and found to be 6.8×10^{-6} g, less than seven millionths of a gram!

The usefulness of spectroscopy in recognizing minute quantities of an element is not new. During the first few years of spectroscopy, five elements that are present on Earth in only tiny concentrations were discovered. For example, the spectrum of minerals found in the water of a certain spring in Germany was analyzed. Two lines of unknown origin were found in the blue region of the spectrum. This bit of evidence was enough to challenge Robert Bunsen, a German chemist, to search for a new element in the water. He evaporated 40,000 kg of spring water to isolate a sample of the new element, which he named "cesium."

In Chapters 5 and 6 you used the characteristic properties of bulk matter to determine the composition of substances. In spectroscopy the properties of atoms are used for the same purpose. The results of the two methods always agree completely.

12. **Figure A shows some spectra observed with the same spectroscope that was used for Figure 8.9. What elements can you identify in (a)? In (b)?**

Figure A
For problem 12

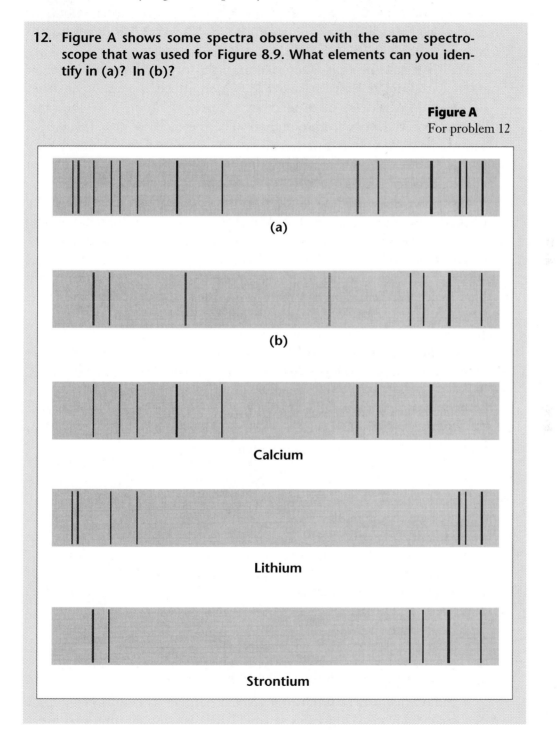

(a)

(b)

Calcium

Lithium

Strontium

"EXPERIMENT"
8.9 An Analog for Radioactive Decay

Having been motivated by radioactivity to develop an atomic model of matter, we are now ready to expand the model with additional assumptions. These assumptions will lead to a testable prediction.

We shall assume that whenever an atom of a radioactive element decays, it emits at least one particle, which affects a counter or other detector. This process takes place by chance: all atoms have the same chance to decay in any time interval. There is no way of telling which atom will decay next. Furthermore, the chance for the decay of a particular atom is not affected by whether other atoms have or have not already decayed.

From these assumptions some startling conclusions can be drawn by using the power of mathematics. We shall not do that. Instead, you will act out the process as directed by the assumptions, as you did in the synthesis of FsR in Section 8.4.

The radioactive atoms will be represented by dice. Each team will begin with 12 dice. The dice will be shaken and thrown on the table. The dice that land with a six on top will be considered to have decayed. Shaking the dice gives each of them the same chance to land with a six on top. Also, how any particular die lands is not affected by the other dice. By carefully watching for the sixes, you will be the counter. Thus, the dice analog satisfies the assumptions of the model.

Each throw stands for some fixed time interval. A record of the number of dice that decay and that remain will illustrate what the model says about the decay of a radioactive element over time.

- Why should you remove the dice with a six on top before you throw the dice again?

- Why should you thoroughly shake the remaining dice before the next throw?

Shake and throw the dice ten times. Be sure to keep a record of the number of dice you started with, the number that decayed, and the number that remained after each throw. Figure 8.11 shows a graph of the number of remaining dice as a function of the number of throws for one run. Draw a graph like Figure 8.11 from your data.

- Is your graph exactly the same as Figure 8.11? Does your graph look exactly like any of the graphs of the teams near you?

A radioactive source producing a small number of counts per minute is considered a weak source. A set of 12 dice is a model for a very weak

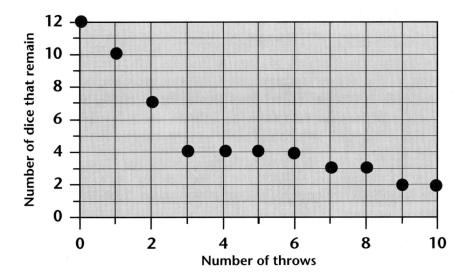

Figure 8.11
A graph of the number of remaining dice as a function of
the number of throws.

radioactive source. To see how a strong radioactive source would act,
you can add the decay data from the entire class. Now you can draw a
smooth curve that passes quite close to the class data.

- How many throws did it take for the number of remaining dice to
  reach half the original value?
- In how many throws did the number of remaining dice drop from
  half the original value to one quarter of the original value?

With dice, you can count both the decaying and the remaining ones.
With a real radioactive element, you can measure only the number of
atoms decaying per unit time, that is, the decay rate. To draw testable
conclusions from the model, you must relate the graph of remaining
dice to a graph of the decay rate.

We assumed that the chance of an atom to decay is independent of
how many atoms are in the sample. Therefore, we expect that the
greater the number of atoms in a sample, the greater the decay rate. By
the same reasoning, a sample with half the number of atoms would have
half the decay rate.

- What do you conclude from the preceding paragraph about the
  shape of a graph of decay rate as a function of the number of
  throws?
- Does your answer to the preceding question remind you of any
  graphs in Chapter 7? If so, which and why?

8.10 Half-Life

The throwing of dice cannot tell us how much time it will take the decay rate of a radioactive element to drop to half its initial value. But the model makes the following prediction for any radioactive element: Whatever time it takes to reduce the decay rate to one half its initial value, it will take the *same* time to reduce it from one half to one quarter of the initial value. It will take the same time to reduce the decay rate from one quarter to one eighth, and so on.

Thus, every radioactive element has a *half-life*, that is, a time during which its decay rate reduces to half its initial value. This half-life is a characteristic property of the element; it is independent of the number of atoms in the sample.

The counting rate measured with a Geiger counter does not equal the decay rate because not all particles enter the counter; some are emitted in other directions. But if we keep the counter always at the same distance from the sample, then the counting rate will be proportional to the decay rate. This prediction can, and has been, tested. All radioactive elements have a half-life. The values of the half-lives for a few elements are given in Table 8.1.

Table 8.1

Element	Half-life	Element	Half-life
Uranium	4.5×10^9 years	Polonium	138 days
Radium	1.6×10^3 years	Radon	3.83 days

13. Recall the assumptions of the model for radioactive decay. Explain in your own words how these assumptions are satisfied by the dice analog.

14. Suppose that at a given moment you have two samples, one of radon and one of polonium. Each sample contains a billion (1×10^9) atoms.
    a. After 3.83 days, which sample will contain more of the original atoms?
    b. Were there more decays of polonium atoms or radon atoms during this period?
    c. Which of the two samples has a higher decay rate?
    d. Under the same conditions, which sample will show a greater counting rate?

15. Suppose you repeat "Experiment" 8.9, An Analog for Radioactive Decay, with the following change: A die showing a five or a three on each throw shall be considered to have decayed.

    a. Will the dice still show a half-life? Why?

    b. If your answer to part (a) was "yes," how will the new half-life relate to the one you observed in Experiment 8.9? Try it, if you want.

FOR REVIEW, APPLICATIONS, AND EXTENSIONS

16. Think again of the sealed can referred to in Section 8.1. You are not allowed to pierce the can or break it open. What would you predict about its behavior in each case?

    a. You lowered the temperature sufficiently.

    b. You raised the temperature sufficiently

17. While doing Experiment 8.2, A Black Box, why were you not permitted to open the box and look inside?

18. Which of the following are *parts* of the atomic model of matter, and which are *illustrations* of the model?

    a. Matter is made up of very tiny particles.

    b. Atoms of the same element are all alike.

    c. A rubber ring combines with a fastener to form FsR.

    d. Atoms of different elements may combine to form compounds.

    e. Marbles can be stacked in a box to represent a solid.

19. In the nineteenth century Louis Pasteur developed a "germ model" for disease. According to Pasteur, little organisms, too small to be seen by the unaided eye, were the cause of disease.

    a. How does this model account for people getting sick?

    b. What predictions does the model make about staying healthy?

    c. What are some of the limitations of this model? (Hint: Think of kidney stones or heart attacks.)

20. Suppose that 6 million atoms of hydrogen combine with 2 million atoms of nitrogen to form 2 million molecules of ammonia. How many atoms of hydrogen and how many atoms of nitrogen would there be in each molecule of ammonia?

21. Refer to the graph of the number of remaining dice as a function of the number of throws ("Experiment" 8.9).

    a. How many throws are needed to reduce the initial number of dice to 80%?

    b. How many throws are needed to reduce the initial number of dice to 80% of 80%, that is, to 64%?

c. How many throws are needed to reduce the initial number to 80% of 80% of 80%, that is, 51%?

d. Is "the time in which 80% of the number of atoms of a radioactive element remain" a characteristic property of the element? If yes, how is this time related to the half-life?

22. Many radioactive elements are waste products in nuclear power plants. Periodically, this waste must be removed and stored in a safe place. Suppose that three radioactive elements are produced. Let their half-lives be 30 minutes, 500 years, and 1×10^8 years. Suppose also that when the waste is ready to be shipped, the three elements are present in equal numbers of atoms.

a. Which of the three elements will have the highest decay rate?

b. The atoms of which of the three elements will still be present in significant numbers after one year?

c. If not stored safely, which of the remaining elements will pose a health hazard after 100 years?

23. a. Since polonium has a half-life of 138 days, it must itself be a product of radioactive decomposition of another element. Why?

b. Will a freshly prepared sample of polonium have the same half-life as the remainder of an old sample?

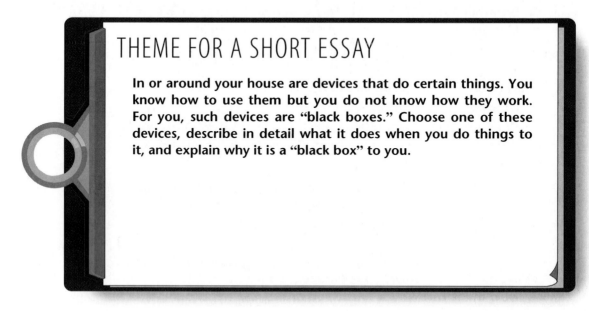

THEME FOR A SHORT ESSAY

In or around your house are devices that do certain things. You know how to use them but you do not know how they work. For you, such devices are "black boxes." Choose one of these devices, describe in detail what it does when you do things to it, and explain why it is a "black box" to you.

Chapter 9

Sizes and Masses of Molecules and Atoms

Molecules and atoms are not visible even under powerful optical microscopes. How, then, can you answer any of the following questions: How big is an atom? What is the mass of one atom? How many atoms are in a sample of matter that you can work with in the laboratory? These three questions are related. If you can measure any one of these quantities, you can calculate the others. You will do just that in this chapter. However, you will use indirect methods. The first method is introduced in the section below.

9.1 The Thickness of a Thin Layer

Consider a rectangular solid. Its volume V is equal to its length times its width times its height, or $V = lwh$ [Figure 9.1(a)]. Since lw is the area of the base, we can say that the volume equals the area of the base times the height.

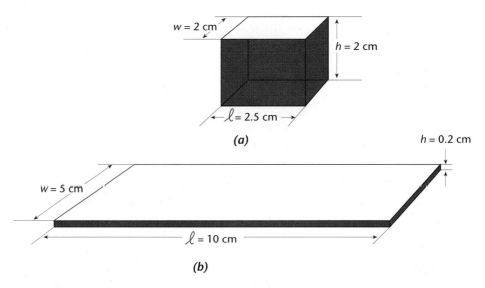

(a)

(b)

Figure 9.1
The piece of matter in (a) has a volume of $lwh = 2.5$ cm $\times$ 2.0 cm $\times$ 2.0 cm = 10 cm^3. In (b) the same piece of matter has been flattened. It has the same volume as in (a). The volume of the thin slab in (b) is $lwh = 10$ cm $\times$ 5 cm $\times$ 0.2 cm = 10 cm^3.

Figure 9.1(b) shows another rectangular solid with the same volume as that in Figure 9.1(a). The material has been formed into a different rectangular shape that has a larger base and a much smaller height. Its volume is still the area of its base times its height. If the material were spread out into an even thinner layer, its volume would be unchanged and would still be equal to the area of its base times its height.

Now suppose you have a thin rectangular sheet of some material—a metal foil, for example—whose volume you know. From measurements of its length and width, you can calculate the area of its base. To find the thickness, you divide the known volume by the area of the base.

EXPERIMENT
9.2 The Thickness of a Thin Sheet of Metal

You can find the thickness of a rectangular sheet of aluminum foil from its length, width, and volume. To find the volume you need to know the mass of the sheet and the density of the material (2.7 g/cm^3).

- From your calculation of the foil's thickness, how thick can the atoms of aluminum be?

- How thin can they be?

Metals do not flatten out into thin layers by themselves; they have to be rolled or hammered. Hammering, however, is a crude process. Even the most skilled goldsmith, making thin gold leaf for lettering on store windows, must stop when the gold leaf is about 10^{-5} cm thick. In such cases, the minimum thickness is determined not by the size of the fundamental particles of gold but by the difficulty of handling such a thin sheet.

Liquids, on the other hand, tend to spread out into thin layers by themselves. If you pour water on the floor, it spreads out quickly to form a thin layer. You can calculate the thickness of this layer if you know the volume of the water you pour out and the area it covers.

†1. In a small-boat harbor, a careless sailor dumps overboard a quart (about 1,000 cm³) of diesel oil. Assume that this oil will spread evenly over the surface of the water to a thickness of 0.00010 cm. What area will be covered with oil?

2. Spherical lead shot are poured into a square tray, 10 cm on a side, until they completely cover the bottom. The shot are poured from the tray into a graduated cylinder, which they fill to the 20-cm³ mark.
   a. What is the diameter of a single shot?
   b. How many shot were in the tray?
   c. If the 20 cm³ of shot had a mass of 130 g, what would be the mass of a single shot?

†3. A tiny drop of mercury has a volume of 0.0010 cm³. The density of mercury is about 14 g/cm³. What is the mass of the drop?

4. A goldsmith hammers 19.3 g of gold into a thin sheet of foil 100 cm in length and 100 cm in width. The density of gold is 19.3 g/cm³.
   a. What is the volume of the gold?
   b. What is the area of the gold sheet?
   c. What is the thickness of the gold sheet?

9.3 Scientific Notation

The number of atoms in a piece of iron is very large. The mass of one atom in grams is a very small number. *Scientific notation* simplifies calculations made with large and small numbers by eliminating the need to use zeros as placeholders. Scientific notation thus enables us to write only the significant digits of a number. The section will help you to use scientific notation even if you did not study it in mathematics class.

Consider the shorthand notation introduced in Table 9.1.

Table 9.1

Words	Standard numerals	Written shorthand	Spoken shorthand
million	1,000,000	10^6	"ten to the sixth"
hundred thousand	100,000	10^5	"ten to the fifth"
ten thousand	10,000	10^4	"ten to the fourth"
thousand	1,000	10^3	"ten cubed"
hundred	100	10^2	"ten squared"
ten	10	(10^1)	(shorthand not used)
one	1	(10^0)	(shorthand not used)
tenth	0.1	10^{-1}	"ten to the negative one"
hundredth	0.01	10^{-2}	"ten to the negative two"
thousandth	0.001	10^{-3}	"ten to the negative three"

The numbers in shorthand in Table 9.1 are called powers of ten. The small numeral at the right of 10 is called the exponent. It is evident from Table 9.1 that the exponent of 10 equals the number of zeros in the standard notation. Numbers larger than 1 have trailing zeros and numbers smaller than 1 have leading zeros. In order to distinguish between the two types of numbers, we write a minus sign before the exponent for numbers smaller than 1.

Any number can be written as a product of a number between 1 and 10, and a power of ten. The following examples show how this is done.

$$3{,}000 = 3 \times 1{,}000 = 3 \times 10^3$$
$$3{,}200 = 3.2 \times 1{,}000 = 3.2 \times 10^3$$
$$0.05 = 5 \times 0.01 = 5 \times 10^{-2}$$
$$0.058 = 5.8 \times 0.01 = 5.8 \times 10^{-2}$$

The number multiplying a power of ten is called the *coefficient*. A number that is written as a product of a coefficient between 1 and 10 and a power of ten is said to be written in scientific notation.

You can think of the process of separating a number into a coefficient and a power of ten in the following way. Moving a decimal point three places to the left divides a number by 1,000; for example,

$$3{,}200 \div 1000 = 3.2.$$

To keep the value of the number unchanged, you have to multiply it by 1,000. Hence $3{,}200 = 3.2 \times 1{,}000$. (In 3,200 the decimal point after the last zero is not written.)

Moving a decimal point two places to the right multiplies a number by 100. To keep the number 0.058 unchanged after moving the decimal point two places to the right, you must divide the new form—5.8—by 100, or multiply it by 0.01. Hence $0.058 = 5.8 \times 10^{-2}$.

5. Write the following numbers as numerals with exponents.
   a. hundred million   b. billion   c. ten thousand
   d. millionth   e. ten millionth   f. billionth
   (Hint: For numbers smaller than one, always include the zero to the left of the decimal point; for example, 0.01, not .01.)

6. Write the following numbers as standard numerals and in words.
   a. 10^7 b. 10^{10} c. 10^{-5} d. 10^{-7}

7. Write the following numbers in scientific notation.
   a. 8,000,000   b. 400,000   c. 370,000
   d. 5,610   e. 423   f. 9,060

8. Write the following numbers in scientific notation.
   a. 0.6   b. 0.07   c. 0.004
   d. 0.0506   e. 0.00042   f. 0.612

9. Write the following numbers as standard numerals.
   a. 3×10^6 b. 4.02×10^8 c. 6.52×10^7
   d. 8×10^{-5} e. 7.2×10^{-3} f. 4.07×10^{-4}

9.4 Multiplying and Dividing in Scientific Notation: Significant Digits

As a preparation for multiplying numbers in scientific notation, let us look first at products of powers of ten:

$$10^3 \times 10^2 = 1{,}000 \times 100 = 100{,}000 = 10^5$$
$$10^6 \times 10^{-4} = 1{,}000{,}000 \times 0.0001 = 100 = 10^2$$
$$10^1 \times 10^{-1} = 10 \times 0.1 = 1 = 10^0$$
$$10^{-2} \times 10^{-5} = 0.01 \times 0.00001 = 0.0000001 = 10^{-7}$$

You can work out other examples, which will lead to the general rule: To multiply powers, add the exponents (with their proper signs!). Thus, in the preceding examples, we could have found the exponents of the products by addition:

$$3 + 2 = 5$$
$$6 + (-4) = 2$$
$$1 + (-1) = 0$$
$$(-2) + (-5) = -7$$

To multiply numbers in scientific notation, change the order of the factors in a product. For example:

$$(3 \times 10^5) \times (2 \times 10^4) = (3 \times 2) \times (10^5 \times 10^4) = 6 \times 10^9$$
$$(1.5 \times 10^{-2}) \times (3.0 \times 10^6) = (1.5 \times 3.0) \times (10^{-2} \times 10^6) = 4.5 \times 10^4$$

In general, to multiply two or more numbers in scientific notation, multiply the coefficients and the powers of ten separately.

Since powers of ten are multiplied by adding their exponents, it is reasonable to expect that powers of ten are divided by subtracting their exponents. The following examples illustrate this rule.

$$\frac{10^7}{10^4} = \frac{10{,}000{,}000}{1000} = 1{,}000 = 10^3 \qquad\qquad 7 - 4 = 3$$

$$\frac{10^{-2}}{10^3} = \frac{0.01}{1{,}000} = 0.00001 = 10^{-5} \qquad\qquad (-2) - 3 = -5$$

$$\frac{10^2}{10^{-2}} = \frac{100}{0.01} = 10{,}000 = 10^4 \qquad\qquad 2 - (-2) = 4$$

$$\frac{10^{-3}}{10^{-2}} = \frac{0.001}{0.01} = 0.1 = 10^{-1} \qquad\qquad (-3) - (-2) = -1$$

To divide numbers in scientific notation, rearrange the numbers in a way similar to that used in multiplication.

$$\frac{6 \times 10^7}{2 \times 10^5} = \frac{6}{2} \times \frac{10^7}{10^5} = 3 \times 10^2$$

$$\frac{4.5 \times 10^{-2}}{1.5 \times 10^{-4}} = \frac{4.5}{1.5} \times \frac{10^{-2}}{10^{-4}} = 3.0 \times 10^2$$

In general, to divide numbers in scientific notation, divide the coefficients and the powers of ten separately.

Note that when two numbers in scientific notation are multiplied or divided, the product or quotient is not automatically given in scientific notation. For example,

$$5 \times 10^2 \times 7 \times 10^4 = 35 \times 10^6$$

Here the coefficient, 35, is not between 1 and 10. However, we can move the decimal point one place to the left, and compensate for this by multiplying the number by 10:

$$35 \times 10^6 = 3.5 \times 10 \times 10^6 = 3.5 \times 10^7$$

Here is another example:

$$\frac{4 \times 10^5}{8 \times 10^2} = 0.5 \times 10^3$$

To change the coefficient to a number between 1 and 10, in this case you move the decimal point one place to the right, and compensate for this by dividing by 10 (or multiplying by 10^{-1}):

$$0.5 \times 10^3 = 5 \times 10^{-1} \times 10^3 = 5 \times 10^2$$

Scientific notation eliminates the need for zeros as placeholders. It removes any doubt about the significance of zeros in a measured number. For example, how many significant digits are there in 3,600 cm^2? Is this measurement accurate to the nearest 100 cm^2, 10 cm^2, or 1 cm^2? If standard numerals are used, there is no way of telling. In scientific notation, however, you write 3.6×10^3 cm^2 to indicate two significant digits, and 3.60×10^3 cm^2 to indicate three significant digits. Thus, you would write 3.60×10^3 cm^2 if the measurement 3,600 cm^2 were reliable to the nearest 10 cm^2.

You are likely to use scientific notation mostly for calculations with measurements. You should therefore use the rule given in Section 3.8 for the number of digits to be written when multiplying or dividing the coefficients: The result should have as many digits as the measurement with the smallest number of digits.

10. Multiply the following numbers.
    a. $10^8 \times 10^6$
    b. $10^9 \times 10^{-5}$
    c. $10^{-2} \times 10^6$
    d. $10^{-3} \times 10^{-3}$
    e. $10^4 \times 10^{-7}$
    f. $10^{-10} \times 10^{-12}$

11. Multiply the following numbers and express the results in scientific notation.
    a. $2 \times 10^7 \times 3 \times 10^{-4}$
    b. $4 \times 10^3 \times 4 \times 10^3$
    c. $5 \times 10^{-2} \times 1 \times 10^{-6}$
    d. $7 \times 10^3 \times 5 \times 10^{-2}$
    e. $2.5 \times 10^4 \times 3 \times 10^5$
    f. $6 \times 10^7 \times 1.5 \times 10^8$

12. Divide the following numbers.
    a. $\dfrac{10^7}{10^3}$
    b. $\dfrac{10^4}{10^8}$
    c. $\dfrac{10^{12}}{10^{-2}}$
    d. $\dfrac{10^{-6}}{10^{-8}}$
    e. $\dfrac{10^{-4}}{10^3}$
    f. $\dfrac{10^{-4}}{10^{-2}}$

13. Divide the following numbers and express the results in scientific notation.
    a. $\dfrac{6 \times 10^5}{4 \times 10^3}$
    b. $\dfrac{2 \times 10^7}{5 \times 10^{-8}}$
    c. $\dfrac{7 \times 10^{-4}}{1 \times 10^{-6}}$
    d. $\dfrac{2 \times 10^8}{8 \times 10^4}$
    e. $\dfrac{3 \times 10^{-2}}{6 \times 10^4}$
    f. $\dfrac{7 \times 10^5}{2 \times 10^2}$

14. Write the following measurements in scientific notation.
    a. 6,500 g to the nearest 100 g
    b. 6,500 g to the nearest 1 g
    c. 6,500 g to the nearest 0.1 g
    d. 200 cm^3 to the nearest 1 cm^3
    e. 5,040 cm^3 to the nearest 10 cm^3
    f. 70,000 cm^3 to the nearest 1,000 cm^3

15. What is the area of a rectangular piece of land with each pair of sides?
    a. 3.5×10^3 m and 1.2×10^3 m
    b. 6.3×10^2 m and 4.7×10^4 m

16. Calculate the mass of 2.0×10^4 cm^3 of oxygen. (See Table 3.1 for the density of oxygen.)

17. Suppose that a 3.2×10^{-3} g sample of a radioactive element produces 4.5×10^4 helium atoms per minute. How many helium atoms would be produced by 1.00 g of the element in 60.0 minutes?

EXPERIMENT
9.5 The Size and Mass of an Oleic Acid Molecule

You are now ready to calculate the size and the mass of a single molecule. The starting point will be finding the thickness of a very thin layer of a substance. To do that, you will follow the same approach that you used in Experiment 9.2, The Thickness of a Thin Sheet of Metal.

Under proper conditions, oil and certain other substances can spread out even more thinly than water. Oil does not dissolve in water, it floats on top of it. You often see oil spread out on a water surface in a thin, rainbow-colored film. Oleic acid, a compound of hydrogen, carbon, and oxygen, does not dissolve in water either but spreads out into an even thinner layer than oil does. In fact, a single drop of this liquid from a medicine dropper will spread so thin that it will cover the entire surface of a small wading pool.

To make it possible to do the experiment in a small tray, we must dilute the solution of oleic acid in alcohol. In this experiment you will be using a solution that is 1 part oleic acid in 500 parts of solution, or 2×10^{-3} parts of oleic acid.

Pour tap water into a tray to a depth of about half a centimeter. Allow the water to stand for a few minutes, until all movement has stopped. Then sprinkle just enough fine powder on the water to make a barely visible layer. When a drop of dilute solution of oleic acid is dropped onto the water surface, the powder will be pushed aside. This allows you to see the oleic acid film when it spreads out across the water (Figure 9.2 on the next page).

To prove that it is the oleic acid and not the alcohol that pushes the powder aside, first put one drop of pure alcohol from a medicine dropper in the center of the tray.

• What do you observe?

Now put one drop of the dilute solution in the center of the tray.

• If the area covered by the film is close to a circle, what is its diameter?

• If the area is only roughly circular, how will you find its average diameter?

• What is the area of the thin film?

To find the volume of the acid in the thin layer, you must first find the volume of one drop of the dilute solution. You can do that by counting the number of drops that will add up to 1.0 cm³ in a graduated cylinder.

• What is the volume of one drop of solution?

Figure 9.2
A close-up of the tray with motionless water. (*a*) After a thin layer of powder has been sprinkled on the water. (*b*) After a drop of dilute solution of oleic acid has been dropped from a medicine dropper.

- What is the volume of oleic acid in one drop of the solution? (Remember that only 1/500 of the solution is oleic acid.)
- From your volume and area measurements, what is the thickness of the layer?*

In your calculation of this fantastically small thickness, you used only the fact that the height of a cylinder equals its volume divided by the area of its base. To draw any conclusions about the size of oleic acid molecules, you have to make the following assumption: The oleic acid spreads out until the layer is only one molecule thick. From this assumption it follows that the thickness you calculated is the height of a single molecule.

So far you have found only the height of a molecule of oleic acid. Other experiments show that oleic acid molecules are long and thin, with a height about 10 times the width of their base. They stand nearly upright on a water surface when they form a thin layer. Thus a tiny piece

---

*The volume of a thin cylinder or disk, like that of a rectangular sheet, is equal to the area of the base times the height, no matter how short the cylinder. In the case of a cylinder, however, the base is a circle and its area is equal to πr^2, where r is the radius of the base.

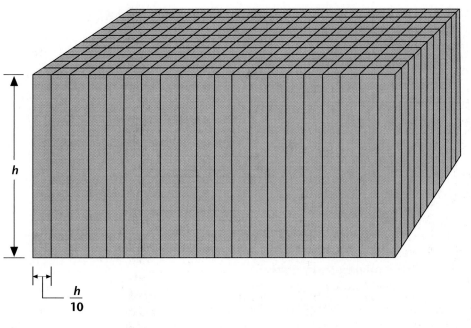

Figure 9.3
A simplified picture of a submicroscopic piece of a film of oleic acid one molecule thick.

of a layer of oleic acid, one molecule thick, might look roughly like the collection of tiny rods in Figure 9.3. The area of the base of each rod is then equal to the width of the base squared, as shown in Figure 9.4.

Suppose that all the molecules of oleic acid touch one another, as in Figure 9.3.

Then,

Area of layer = (number of molecules) · (base area of one molecule)

or,

$$\text{Number of molecules} = \frac{\text{area of layer}}{\text{base area of one molecule}}$$

- From your data for the height of one molecule and the fact that the width is about 1/10 of the height, what is the area of the base of one molecule?
- What is the number of molecules in the layer?
- What does this tell you about the number of molecules in the droplet you started with?

From the volume of the oleic acid in one drop of dilute solution and the density of oleic acid, you can calculate the mass of oleic acid in one drop of dilute solution. The density of oleic acid is 0.87 g/cm^3. Since this

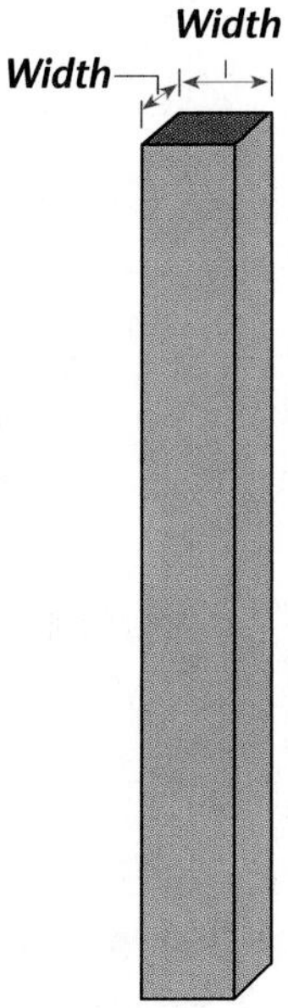

Figure 9.4
An enlarged view of a simplified picture of a single oleic acid molecule. The area of the base (the darker region) is equal to (width)^2.

is a very rough calculation, you may substitute 1 g/cm³ for the density of oleic acid.

- What is the mass of the droplet of oleic acid?

Knowing the number of molecules in the droplet, you can now use the relation

$$\text{Mass of one molecule} = \frac{\text{mass of sample}}{\text{number of molecules in sample}}$$

to find the mass of a single molecule of oleic acid.

- What is the mass of one molecule of oleic acid?

18. a. What is the volume of oleic acid in 1.0 cm³ of solution prepared by dissolving 2.0 cm³ of oleic acid in 98 cm³ of alcohol and mixing thoroughly?
     b. If 50 drops of this solution occupy 1.0 cm³, what is the volume of one drop of solution?
     c. What is the volume of oleic acid in one drop of solution?

19. If 3×10^{-5} cm³ of pure oleic acid forms a film with an area of 150 cm², how thick is the film?

9.6 The Mass of Helium Atoms

By indirect but simple means, you were able to determine the mass of a single molecule of oleic acid. Now we shall describe an experiment in which the mass of a single helium atom is found. The experiment was done for *IPS* students at the Mound Laboratory of the Monsanto Research Corporation.

As in your experiment with oleic acid, an indirect method was used. The main idea of the experiment was to prepare a sample of helium of known mass and known number of atoms. The mass of a single atom was then calculated by division:

$$\text{Mass of atom} = \frac{\text{mass of sample}}{\text{number of atoms in sample}}$$

Figure 9.5
The sealed quartz tube containing polonium. The polonium produces a blue glow, caused by the emitted helium particles when they strike the quartz. This photograph was taken in the light from the blue glow.

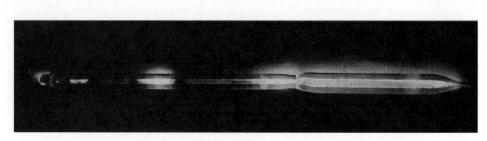

The helium was produced by the radioactive decay of polonium. A small amount of polonium was placed inside a quartz tube of known diameter. The air was pumped out and the tube was sealed (Figure 9.5 on the previous page).

After 21 days the seal was broken underwater. The water rushed into the tube and compressed the helium to atmospheric pressure (Figure 9.6). From the volume of the helium and its known density at atmospheric pressure, the mass of the helium sample was calculated. The number of helium atoms was calculated from the average counting rate per minute and the number of minutes in 21 days.

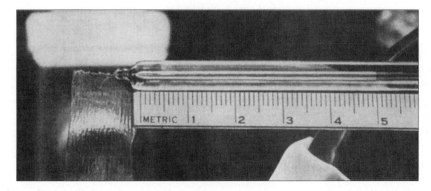

Figure 9.6
The quartz tube containing helium after being opened underwater. The length of the gas column is 5.0 cm.

In principle, the polonium could now be removed from the tube and placed in a counter in order to find the number of disintegrations per minute. However, in practice this is not possible. A sample large enough to produce a measurable amount of helium in three weeks would give off too many helium particles per minute for the counter to count.

Therefore, the polonium was first dissolved completely in a large quantity of nitric acid. The solution was further diluted with water. In both steps, the solution was thoroughly mixed to make sure the polonium was evenly distributed throughout. Finally, a tiny drop of the solution was put on a small plate, the acid was evaporated, and the plate placed in a counter (Figure 9.7).

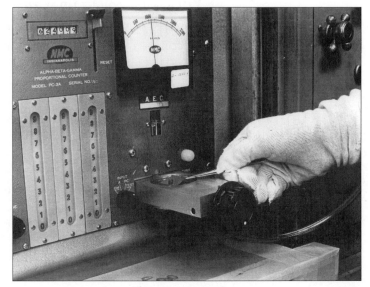

Figure 9.7
A metal plate containing polonium being removed from a radiation counter after a count.

Here is the record of the data of the experiment:

Production of Helium

On March 3 the polonium was sealed in the evacuated quartz tube. The seal was broken on March 24. The water rose in the tube and the helium was compressed to a length of 5.0 cm in the tube (Figure 9.6).

The area of the cross section of the inside of the quartz tube, checked beforehand, was 8.1×10^{-3} cm^2. Thus the volume of helium in the tube was

$$\begin{aligned} \text{Volume} &= \text{length} \cdot \text{area of cross section} \\ &= 5.0 \text{ cm} \times 8.1 \times 10^{-3} \text{ cm}^2 \\ &= 4.1 \times 10^{-2} \text{ cm}^3 \, . \end{aligned}$$

The density of helium at atmospheric pressure and room temperature is 1.7×10^{-4} g/cm^3. The mass of the sample of helium is

$$\begin{aligned} \text{Mass} &= \text{volume} \cdot \text{density} \\ &= 4.1 \times 10^{-2} \text{ cm}^3 \times 1.7 \times 10^{-4} \text{ g/cm}^3 \\ &= 7.0 \times 10^{-6} \text{ g}. \end{aligned}$$

Dilution

The small sample of polonium was first dissolved in 1.0×10^3 cm^3 of nitric acid. Therefore, one cubic centimeter of this solution contains 1/1,000 of the original amount of polonium.

A volume of 1.0 cm^3 of this solution was then mixed with 99 cm^3 of water. Hence 1.0 cm^3 of this dilute solution contained only 1 part in 10^5 (or 1.0×10^5) of the original sample of polonium. Even this was too much to be counted!

Therefore, only 1.0×10^{-3} cm^3 (one one-thousandth of a cubic centimeter) of the very dilute polonium solution was placed on the plate to be counted. This meant that only

$$1.0 \times 10^{-5} \times 1.0 \times 10^{-3} = 1.0 \times 10^{-8},$$

or 1 part in 100 million, of the original sample of polonium was counted. Two trials were run with samples of this size.

Counting

$$\text{Trial 1: } 2.4 \times 10^5 \text{ counts/minute}$$
$$\text{Trial 2: } 2.0 \times 10^5 \text{ counts/minute}$$

This step introduces the largest experimental error on the experiment so far. The average of the two readings, 2.2×10^5 counts/minute, was used.

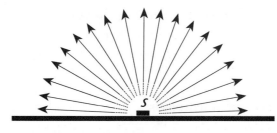

Figure 9.8
Helium particles flying off from a thin polonium source S on a metal plate P. Those emitted in the directions shown by the arrows can be counted. Half of the helium particles are emitted downward into the plate and are not counted.

The counter used in the experiment described above does not count those helium particles which fly off into the plate; it counts only those which fly off upward (Figure 9.8). Thus the true disintegration rate was twice the recorded number, 4.4×10^5 counts/minute.

This large number of counts is produced by only 1.0×10^{-8} of the original sample. The number of disintegrations per minute of the whole sample is

$$1.0 \times 10^8 \times 4.4 \times 10^5 \text{ counts/min} = 4.4 \times 10^{13} \text{ counts/min.}$$

Now recall that the quartz tube containing the polonium was sealed on March 3 and opened on March 24. This is a period of 21 days or

$$21 \text{ days} \times 24 \text{ hours/day} \times 60 \text{ min/hour} = 3.0 \times 10^4 \text{ min.}$$

If the sample of polonium had decayed at the same rate over the entire duration of the experiment, then the total number of counts would have been

$$4.4 \times 10^{13} \text{ counts/min} \times 3.0 \times 10^4 \text{ min} = 1.3 \times 10^{18} \text{ counts.}$$

Of course, the counting rate is not constant; it decreases with time. In fact, after 138 days it is down to half its original value (Figure 9.9). After 21 days the rate of decay is 0.90 of the original rate. So the average rate of decay over the first 21 days is about 0.95 of the original rate. Thus the counting rate measured after 21 days must be divided by 0.95 to give the average counting rate over the 21 days. This correction will raise the total number of counts during the experiment to 1.4×10^{18}.

Therefore, we shall use the value 1.4×10^{18} for the total number of counts during the experiment.

We said earlier that we shall assume that each decay signals the formation of one helium atom. Thus the number of helium atoms produced by the polonium sample during the three weeks is:

$$\text{Number of helium atoms} = 1.4 \times 10^{18}$$

and the mass of one helium atom is

$$\text{Mass of atom} = \frac{\text{mass of sample}}{\text{number of atoms in sample}}$$

$$= \frac{7.0 \times 10^{-6} \text{ g}}{1.4 \times 10^{18}} = 5.0 \times 10^{-24} \text{ g}$$

20. If the mass of an atom of an element is 5.0×10^{-23} g, how many atoms are there in 1 g of that element?

†21. A cylindrical tube with a cross-sectional area of 2 cm² and a height of 50 cm is filled with hydrogen. What is the volume of the hydrogen in the tube?

22. Write a brief summary of the steps followed in finding the mass of helium atoms by radioactive decay.

†23. In the experiment on the mass of helium, how many lead atoms were formed? What assumptions are made to get this number?

9.7 The Mass of Polonium Atoms

We can apply the relation

$$\text{Mass of atom} = \frac{\text{mass of sample}}{\text{number of atoms in sample}}$$

to polonium as well as to helium. We assumed that one atom of polonium disintegrates into one atom of lead and one atom of helium. Therefore, the number of atoms of polonium that disintegrated during the three weeks of the experiment equals the number of helium atoms formed. This number we calculated in the preceding section. To calculate the mass of one polonium atom, we must know the mass of the polonium that disintegrated.

As we saw in the preceding section, the rate of decay of polonium after 21 days is 0.90, or 90%, of the original rate. Since the rate of decay is proportional to the number of polonium atoms present, 0.90 of the original mass of polonium remained in the quartz tube. In other words, 0.10 of the initial sample had decayed.

All this we learn from Figure 9.9. But to find out how much polonium decayed, you have to know the initial mass of the polonium. Although we did not mention it before, the polonium was massed before it was placed in the quartz tube (see Figure 9.10 on the next page).

Massing polonium is difficult. Because of the intense and dangerous radiation, the polonium must be massed inside a small closed container. The procedure resembles the massing of a liquid. Here we state only the final result, found by massing the container plus polonium, then subtracting the mass of the container.

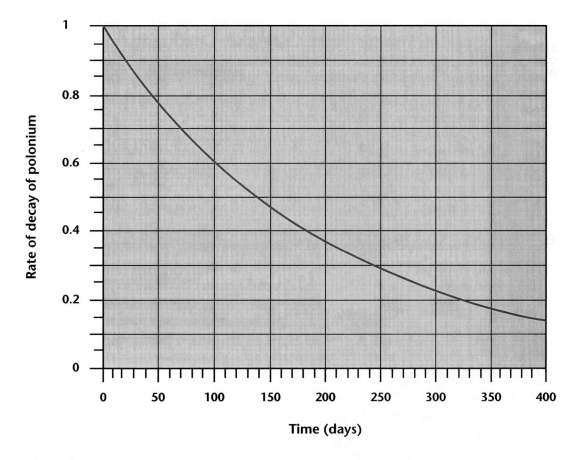

Figure 9.9
The rate of decay of a sample containing polonium as a function of time. The rate (number of polonium atoms that disintegrate per unit time) is expressed in terms of the fraction of the rate measured at zero days. The rate reaches 0.5 of the original value at 138 days, the half-life of polonium.

$$\text{Mass of polonium sample} = 4.5 \times 10^{-3} \text{ g}$$

$$\text{Mass of polonium that decayed} = 0.10 \times 4.5 \times 10^{-3} \text{ g}$$

$$= 4.5 \times 10^{-4} \text{ g}$$

$$\text{Number of atoms of polonium that decayed} = 1.4 \times 10^{18}$$

$$\text{Mass of one polonium atom} = \frac{4.5 \times 10^{-4}}{1.4 \times 10^{18}} = 3.2 \times 10^{-22} \text{ g}$$

$$= 320 \times 10^{-24} \text{ g}$$

We wrote the mass of a polonium atom to the same power of ten as for a helium atom. We see that polonium atoms are about $320/5.0 = 64$ times as heavy as helium atoms.

The two experiments involve the determination of the mass and number of atoms in a small sample. They are by no means the first or the most accurate of such determinations, but they are perhaps the most direct. A repetition of this experiment, which you may see in an *IPS* video, gave a similar result: 7.5×10^{-24} g for helium and 410×10^{-24} g for polonium. For higher precision, different methods are used and give a mass of 6.64×10^{-24} g for helium and 349×10^{-24} g for polonium.

Before studying this experiment, you had no way of guessing what the mass of a single atom might be. The mass might have been a billion times larger than 10^{-24} g—that is, about 10^{-15} g—or perhaps 10^{-33} g, a billion times smaller. Thus, being able to conclude that the mass of a helium atom is close to 6×10^{-24} g is an impressive achievement.

Although the masses of atoms vary over a wide range, they are all very small compared with the masses you are likely to put on a balance. It would require frequent writing of factors of 10^{-24} if we expressed

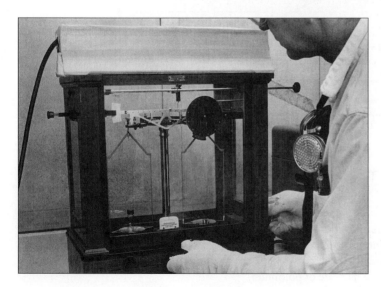

Figure 9.10
A few milligrams of polonium being weighed on a sensitive balance. The polonium is in a small glass tube in the sealed bottle on the left-hand pan. The bottle is sealed to prevent contamination of the balance and the laboratory by the polonium.

masses of atoms in grams. It is therefore convenient to choose the mass of the lightest known atom, that of a hydrogen atom, as the *unified atomic mass unit* (u). Table 9.2 lists the atomic masses of some common elements in both grams and unified atomic mass units.*

Table 9.2

atom	Mass (g)	(u)	atom	Mass (g)	(u)
Aluminum	4.48×10^{-23}	27.0	Mercury	3.34×10^{-22}	201
Calcium	6.66×10^{-23}	40.1	Nickel	9.75×10^{-23}	58.7
Carbon	1.99×10^{-23}	12.0	Nitrogen	2.33×10^{-23}	14.0
Chlorine	5.89×10^{-23}	35.5	Oxygen	2.66×10^{-23}	16.0
Copper	1.05×10^{-22}	63.5	Phosphorus	5.15×10^{-23}	31.0
Gold	3.27×10^{-22}	197	Potassium	6.49×10^{-23}	39.1
Helium	6.64×10^{-24}	4.00	Sodium	3.82×10^{-23}	23.0
Hydrogen	1.66×10^{-24}	1.01	Strontium	1.45×10^{-22}	87.6
Iodine	2.11×10^{-22}	127	Sulfur	5.33×10^{-23}	32.1
Iron	9.27×10^{-23}	55.8	Thorium	3.85×10^{-22}	232
Lead	3.44×10^{-22}	207	Tin	1.98×10^{-23}	119
Lithium	1.15×10^{-23}	6.94	Uranium	3.95×10^{-22}	238
Magnesium	4.04×10^{-23}	24.3	Zinc	1.09×10^{-22}	65.4

24. How many atoms of aluminum are there in 1.0×10^{-3} g of aluminum?

25. Suppose that the smallest mass that you can measure on your balance is 0.005 g. What is the smallest number of carbon atoms that you could mass on your balance?

26. Write a brief summary of the steps followed in finding the mass of polonium atoms by radioactive decay.

9.8 The Size of Atoms

Earlier in this chapter you found the size of a molecule of oleic acid. But a molecule of oleic acid is made up of many atoms. What is the size of a single atom? How can it be found?

---

*Today a unified atomic mass unit is defined differently, but the mass of a hydrogen atom, using the modern definition, is very close to 1 u. The masses given in Table 9.2 are based on the modern definition of a unified atomic mass unit.

To find the mass of a single atom, you used the relation

$$\text{Mass of atom} = \frac{\text{mass of sample}}{\text{number of atoms in sample}}$$

Can you use the relation

$$\text{Volume of one atom} = \frac{\text{volume of sample}}{\text{number of atoms in sample}}$$

to find the volume of one atom? If you can find the volume of an atom and assume a specific shape for it, you can find its size.

To answer the question above, consider the following analogy. Calculating the volume of one student by dividing the volume of the classroom by the number of students will surely be misleading. There is plenty of empty space between any two students. However, finding the volume of one person by dividing the volume of an elevator by the number of people packed tightly in it, may give more reliable results.

In Section 8.5 the incompressibility of solids was related to the idea that in solids, atoms touch each other. Therefore, for solids the above relation makes sense. We shall apply this relation to polonium.

In our experiment, 4.5×10^{-4} g of polonium decayed. The density of polonium is 9.4 g/cm^3. Thus the volume of the polonium sample that decayed was:

$$\text{Volume of sample} = \frac{4.5 \times 10^{-4}\,\text{g}}{9.4\,\text{g/cm}^3} = 4.8 \times 10^{-5}\,\text{cm}^3.$$

The sample of helium produced in our experiment was, of course, a gas. To be able to relate the volume of one helium atom to the volume of the sample, we must find the volume of the sample as a liquid. The atoms of a liquid may be considered to be touching.

There were 7.0×10^{-6} g of helium produced in the experiment. The density of liquid helium (see Table 3.1) is 0.15 g/cm^3. Thus, if we had liquefied our sample of helium it would have had a volume of:

$$\text{Volume of liquid helium} = \frac{7.0 \times 10^{-6}\,\text{g}}{0.15\,\text{g/cm}^3} = 4.7 \times 10^{-5}\,\text{cm}^3$$

This is nearly equal to the volume we found for the sample of polonium. Since each atom of polonium produces one atom of helium, the number of helium atoms produced equals the number of polonium atoms that decayed. In our experiment this number was 1.4×10^{18}. Therefore, the volume of one atom of polonium or helium is about:

$$\text{Volume of one atom} = \frac{4.8 \times 10^{-5}\,\text{cm}^3}{1.4 \times 10^{18}} = 3.4 \times 10^{-23}\,\text{cm}^3.$$

Different data might result in volumes that differ, but not by much.

If we think of atoms as tiny cubes, then the length of the edge of one cube would be

$$\sqrt[3]{3.4 \times 10^{-23} \text{ cm}^3} = 3.2 \times 10^{-8} \text{ cm}.$$

Suppose we think of atoms as tiny spheres enclosed in cubes. Then the diameter of each sphere will be equal to the length of the edge of the cube (Figure 9.11). The diameters of atoms of several elements are shown in Table 9.3.

None of the experiments that you have done in this course suggest any particular shape for atoms. The values of the diameters in Table 9.3 should be considered only as a general indication of size. As such, the table is truly remarkable. The masses of atoms (first column) vary over a factor of about 200, but their diameters (second column) vary hardly at all. Unlike people, heavier atoms are not necessarily bigger! This observation raises new questions about the structure of the atoms themselves.

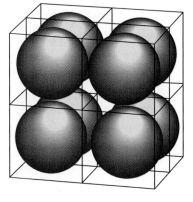

Figure 9.11
Spheres touching all faces of the cubes containing them. The diameter of the spheres equals the length of the edge of the cubes.

Table 9.3

Element	Mass of atom (u)	Diameter of atom (cm)
Aluminum	27.0	2.6×10^{-8}
Bromine (liquid)	79.9	3.5×10^{-8}
Copper	63.5	2.3×10^{-8}
Helium (liquid)	4.00	3.5×10^{-8}
Hydrogen (liquid)	1.01	2.9×10^{-8}
Gold	197.0	2.6×10^{-8}
Lithium	6.94	2.8×10^{-8}
Platinum	195.0	2.5×10^{-8}

FOR REVIEW, APPLICATIONS, AND EXTENSIONS

27. A rectangular object 3.0 cm × 4.0 cm × 5.0 cm is made of many tiny cubes, each 1.0×10^{-2} cm on an edge. How many cubes does the object contain?

28. The diameter of a tennis ball is about 0.07 m, and the dimensions of a tennis court are about 10 m × 24 m. How many tennis balls will be required to cover the court?

29. If the molecules of the oleic acid layer you made could be placed end to end in a line, about how long would it be?

30. If a 10^{-3}-g sample of radium produces 4×10^7 counts/min, how much radium would give 100 counts/min?

31. Suppose you buy a 2-kg bag of dried beans and you find it has been contaminated with small stones. How would you go about finding the approximate number of stones in the bag without separating all the stones in the bag from the beans? What assumptions have you made?

32. A cubic millimeter (10^{-3} cm^3) of blood is found to contain about 5×10^6 red blood cells. The volume of blood in an adult human body is about 5×10^3 cm^3. About how many red blood cells are there in an adult human body?

33. Some polonium is dissolved in 1,000 cm^3 of nitric acid, and a 0.01-cm^3 sample of the solution is counted. Then the number of disintegrations per minute is found to be 3×10^3. How many disintegrations per minute occurred in the original solution?

34. If 10^{18} atoms of polonium disintegrate to produce lead and 10^{-5} g of helium, what is the mass of a helium atom?

35. a. What fraction of a sample of polonium will decay in 100 days? (See Figure 9.9.)
    b. If a counter initially records 5×10^4 counts/min for the sample, what would you expect it to record after 100 days?

36. What value would you find for the volume of one atom of helium if you calculated it from the experimental data in the chapter and the equation below?

$$\text{Volume of one atom} = \frac{\text{volume of gas sample}}{\text{number of atoms in gas sample}}$$

37. A penny is about 1 mm thick. About how many layers of copper atoms does it contain?

THEMES FOR SHORT ESSAYS

1. A few grains of sand can be placed on an overhead projector and projected on the wall. The images are large enough to be measured with a ruler. To be able to calculate the size of the grains of sand, you also need to project an object of known size. Do this investigation and find the size of one sand grain. Then, using the results of your own experimentation on the volume of sand (Experiment 1.4), write a research report, "The Number of Sand Grains in a Cup of Sand."

2. Write an experimental section for an *IBS (Introductory Biological Science)* textbook to count the hairs on a person's head.

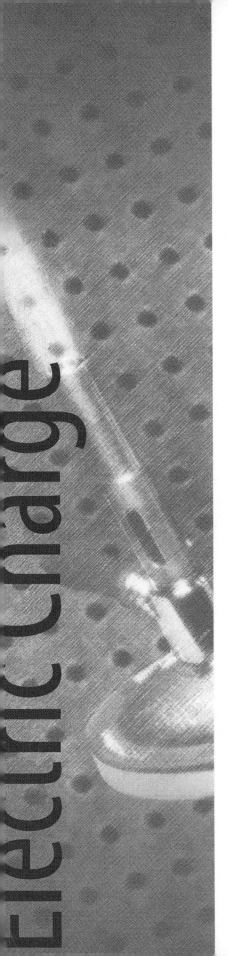

Chapter 10
Electric Charge

10.1 Introduction

In Chapter 6 you used electricity to decompose water. Now we shall come back to this use and examine the connection between electricity and matter.

Figure 10.1 shows the apparatus used in Chapter 6 for the decomposition of water. With only one of the electrodes connected to the battery, no water decomposes. Nothing happens unless both electrodes are connected to the terminals of the battery.

Any household electrical appliance—be it a light bulb, a motor, or a television—has two contacts that have to be plugged in for the device to operate. This common characteristic gave rise in the eighteenth century to the idea that when an electrical device is working, something is moving through it. That something is called electric charge. When you pull out a plug, turn off a switch, or disconnect a battery, the flow of electric charge stops, and with it the operation of the apparatus.

The idea of a flowing electric charge is quite attractive, because it permits us to create a mental picture that may eventually lead to a useful model. To develop the intuitive idea of a flowing electric charge into a model, we must find a way to measure the quantity of electric charge that flows through a bulb, a motor, or any other device.

This situation resembles the one we encountered at the beginning of this course. We felt that there was more substance in a rock than in a pebble, but we needed the balance to make a quantitative comparison. However, while we can see the rock and the pebble directly, we cannot see electric charge, either at rest or in motion. We have to look for an effect produced by moving charge that can be measured quantitatively.

You have used indirect methods of measurement many times before, probably without noticing it. For example, you cannot see temperature directly. To measure temperature, we use the fact that substances expand when heated. We construct various kinds of thermometers using thermal expansion. We shall use a similar method to build a charge meter.

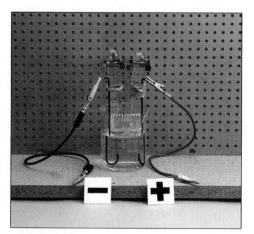

Figure 10.1
Apparatus used for the decomposition of water in Chapter 6.

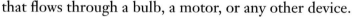

†1. **Which of the following quantities did you measure directly and which did you measure indirectly?**

    a. **Density of a solid in Experiment 3.9.**

    b. **Mass of metal cube in Experiment 3.9.**

    c. **Density of a gas in Experiment 3.11.**

2. **What do you think an electric switch does?**

3. **Some lights used on trucks and automobiles are mounted by means of a metal bracket on the vehicle body. These lights have only one wire, instead of two, to be connected to a switch and battery. How do you explain the operation of such a light?**

10.2 A Measure for the Quantity of Charge

In the experiment on the decomposition of water (Section 6.2), you noted that the longer the electrodes are connected to the battery, the greater the volume of both gases produced. This suggests that more charge flowed through the apparatus when it was connected for a longer time. Thus it seems reasonable to use the quantity of either gas produced in the reaction as a measure of the quantity of electric charge that passes through the water.

We shall choose the quantity of hydrogen as the measure of the quantity of charge. Since we get twice as much of this gas as we do of oxygen, it is easier to detect small quantities of charge. This apparatus, which you will use as a charge meter, we shall refer to as a "hydrogen cell." It is constructed as shown in Figure 10.2. Notice that since you will not be measuring the amount of oxygen, there is no provision to collect it.

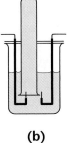

(b)

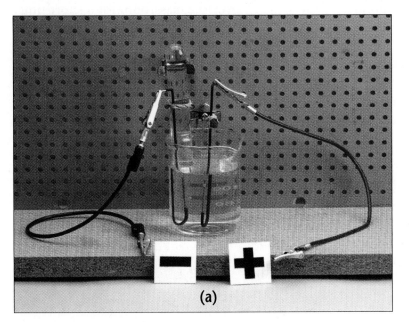

(a)

Figure 10.2
(a) Apparatus for decomposing water. The electrode at the right, at which oxygen is given off, is outside the test tube as shown in (b), so that only hydrogen is collected.

A source of electricity, such as a battery or a wall outlet, and one or more electrical devices connected to the source make up what is called an electric circuit. If you want to know how much charge flows through a given part of an electric circuit [Figure 10.3 (a) is an example], you break the circuit at that place and insert the hydrogen cell [Figure 10.3 (b)]. The amount of hydrogen collected measures how much charge passed through the cell.

Figure 10.3
To measure the charge that flows past point X in the circuit shown in (a), the circuit is broken and a hydrogen cell is inserted as shown in (b).

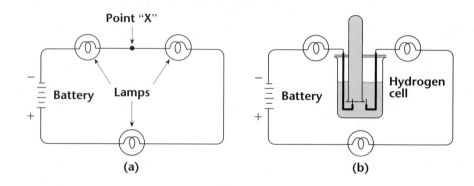

(a) (b)

You will recall that volume is not a reliable measure of the quantity of matter. This is particularly true for a gas, since a gas expands and contracts appreciably as the pressure and the temperature change. Thus, to be accurate, we should measure the quantity of electric charge in terms of the mass, rather than the volume, of hydrogen collected in the test tube.

However, we can be quite sure that the temperature and pressure of the hydrogen are nearly the same all over the classroom for a short time. Therefore, we can be satisfied with simply comparing the volumes of hydrogen collected in the test tubes of different hydrogen cells at about the same time.

We can choose any convenient volume of hydrogen in a test tube as our unit of electric charge. We shall use for our unit the charge needed to produce 1.0 cm^3 of hydrogen.

---

†4. An electrolytic cell for producing hydrogen and oxygen is allowed to run for 5 minutes, and then the battery terminals are reversed. It now runs for an additional 5 minutes.

   a. Does the same quantity of charge flow in each 5-minute interval?

   b. What is the ratio of the volume of gas in one tube to the volume of gas in the other?

EXPERIMENT
10.3 Hydrogen Cells and Light Bulbs

If charge flows around the electric circuit in Figure 10.4, how will the volumes of hydrogen that will collect in the two hydrogen cells compare? Check your prediction by connecting two hydrogen cells, a battery, and a flashlight bulb as shown.

In order to collect hydrogen and not oxygen, be sure that the electrode under the test tube in each of the cells is connected to the wire

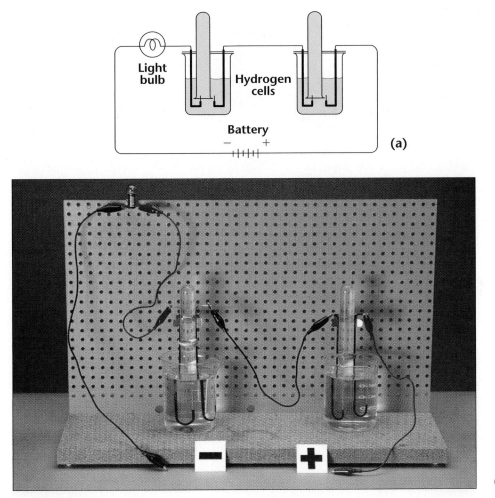

Figure 10.4

(a) A diagram of two hydrogen cells and a light bulb connected in series to a battery. (b) A photograph of the cells ready to be connected to a battery or power supply. The wire at the left is to be connected to the negative (–) terminal; the wire at the right is to be connected to the positive (+) terminal. Note that the negative terminal is connected, through the light bulb, to the hydrogen-producing electrode in the left-hand cell. The wire from the other electrode of this cell goes to the hydrogen-producing electrode of the right-hand cell.

that leads to the negative (–) terminal of the battery. Using a battery of eight flashlight cells, collect hydrogen gas until the water level has dropped about 10 cm in one of the tubes.

After disconnecting the battery, mark the water level in each tube with a small rubber band or a grease pencil. You can use a graduated cylinder to measure the volume of gas collected in each test tube.

• Do your results agree with your prediction?

Now, rearrange the apparatus so that you can measure the charge that flows into and out of the light bulb (Figure 10.5).

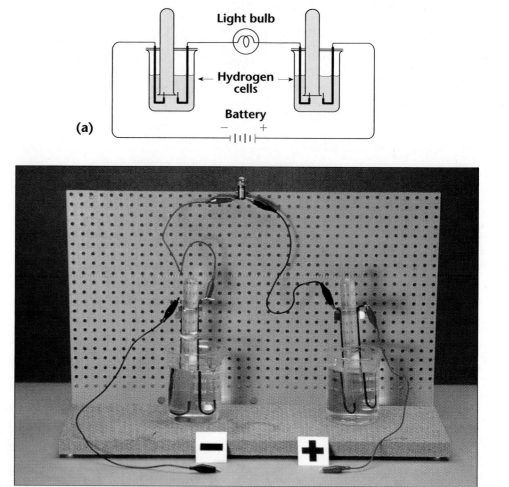

Figure 10.5
(a) The same circuit as shown in Figure 10.4 (a), except that the light bulb is now between the two hydrogen cells. This allows both the charge entering the bulb and that leaving the bulb to be measured. (b) A photograph of the connections to the apparatus.

- How will the gas volumes in the two hydrogen cells compare in this circuit?
- Do your results confirm your prediction?

EXPERIMENT
10.4 Flow of Charge at Different Points in a Circuit

The circuits shown in Figures 10.6 and 10.7 (pages 200 and 201) are slightly more complex than the series circuits shown in Figure 10.4. They are branched or parallel circuits with the light bulb inserted at different points.

Investigate the charge flowing in each of the circuits. Again, in each case allow charge to flow until the liquid level in the test tube in the cell on the right falls about 10 cm.

- What do you predict the liquid levels will be in the other tubes?
- What is the relationship between the amounts of charge flowing in different parts of the circuit?

5. A student connects a battery to a series circuit containing a hydrogen cell followed by a light bulb that connects to another hydrogen cell. The student obtains twice as much gas in one test tube as in the other. How can this be explained?

6. Each of two students connects the circuit shown in Figure 10.6(a) and each collects gas for the same time interval. One of the students is unaware of the fact that the connections to the battery are the reverse of those shown in Figure 10.6(a). How would the charge measured, in cm³ of gas, compare with the charge measured by the student who connected the circuit correctly?

7. Two identical light bulbs are connected to a battery as shown in Figure A. How does the charge that flows past point A in one minute compare with that flowing past points B and C in the same time interval?

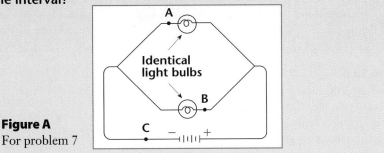

Figure A
For problem 7

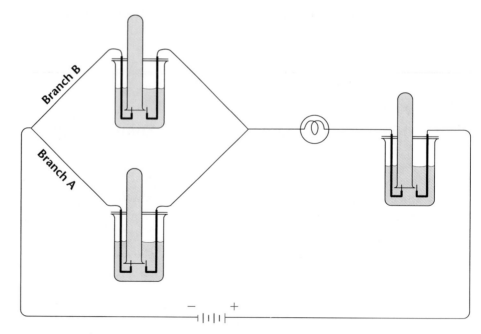

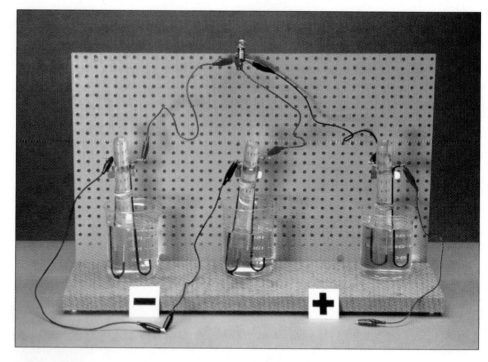

Figure 10.6
(a) Two hydrogen cells that are connected in parallel are connected to a light bulb and a third hydrogen cell that are connected in series. (b) Apparatus connected according to the circuit in (a). The two hydrogen cells connected in parallel are on the left.

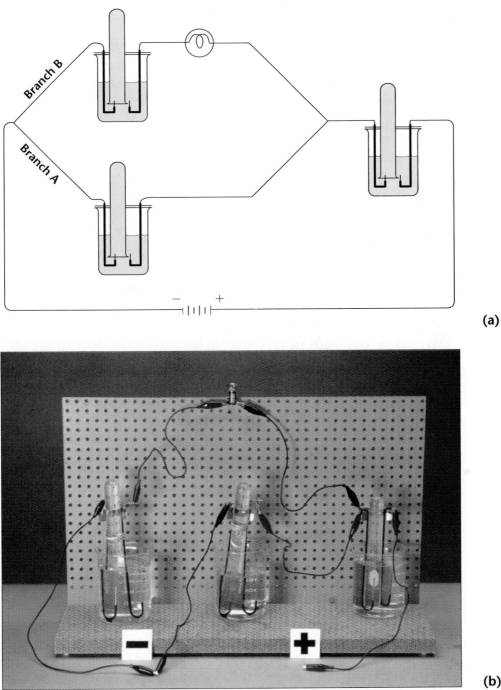

(a)

(b)

Figure 10.7
(a) The same circuit as in Figure 10.6(a), but with the light bulb in one of the branches of the parallel circuit. (b) The apparatus connected according to the circuit in (a). The light bulb and the hydrogen cell on the extreme left are connected in series and make up branch B of the parallel circuit.

10.5 The Conservation of Electric Charge

In the preceding section you examined the quantity of charge that passed through different parts of several electric circuits during a given time. The circuits were rather simple ones; besides the battery, hydrogen cells, and some wire, they contained only a small electric bulb. In all cases the results were consistent with this idea: the same amount of charge passed through all devices connected in series and no charge was consumed.

What about a more complicated circuit—for example, one that also contains an electric motor and two radios (Figure 10.8)? Numerous circuits containing various kinds of electrical devices have been examined, always with the same results. As long as all devices are connected in series, one after another, the same quantity of charge passes through each device. When two or more wires branch off from one point, the sum of the charges passing through all parallel sections equals the charge that flows through the wire before the branch point. In this case as well, no electric charge is lost and none is created.

Many different kinds of experiments have added support to the idea that in an electric circuit no charge is created and no charge is destroyed. The results of these experiments resemble those of the experiments investigating the change in mass in various processes (Sections 2.1 and 2.4–2.6). Within the accuracy of the measurements, your own as well as those of others, we concluded that the total mass did not change in these reactions. We generalized these results into a law, the law of conservation of mass.

Within the accuracy of the measurements, electric charge is neither created nor destroyed when it flows around a circuit. This fact suggests a similar generalization. It is known as the law of conservation of electric charge: Electric charge is never destroyed or created.

By now we have great confidence in the law of conservation of charge. When you encounter a situation where the law seems to be violated, you should check very carefully for some ways by which charge may have flowed unnoticed. For example, suppose a wire carrying electric charge is supported by poorly insulating materials. Some charge will leak through

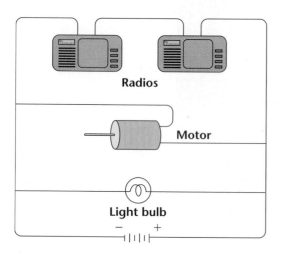

Radios

Motor

Light bulb

Figure 10.8

A circuit consisting of two radios in series connected in parallel with an electric motor and a light bulb.

the insulator and return to the battery through a different path. If this is not taken into account, this situation will look like a violation of charge conservation. When the leaking charge is measured, conservation is found to hold.

8. Would the bulb in Figure 10.6 have glowed with more, less, or the same brightness if it had been placed between the battery and the hydrogen cell on the right?

†9. Three hydrogen cells are connected to a battery (Figure B). When 30 cm³ of hydrogen is collected in tube 1, tube 2 contains 20 cm³ of hydrogen. What is the volume of hydrogen in tube 3 at this time?

Figure B
For problem 9

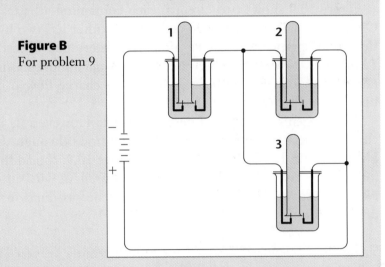

10. Identical hydrogen cells are inserted in a circuit at the points marked "X" in Figure C. Compare the volumes of hydrogen gas that would be collected in the cells at A, D, and F in equal times.

Figure C
For problem 10

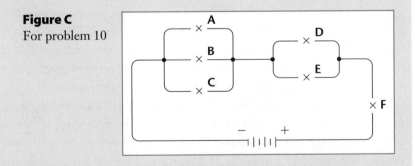

11. In Experiment 10.4, Flow of Charge at Different Points in a Circuit, suppose that some gas escaped unnoticed while gas was collected in the test tube in branch A of Figure 10.7(a). Would the results suggest that charge was created or that it was destroyed?

10.6 The Effect of the Charge Meter on the Circuit

You have seen that two hydrogen cells connected in series give the same readings, that is, a hydrogen cell does not consume charge. Does this mean that such cells do not affect the circuit at all?

Figure 10.9 shows two circuits, each containing a battery and a bulb. One circuit has one hydrogen cell and the other has two hydrogen cells. Both photographs were taken after the battery had been connected for 10 minutes. Notice that the water level dropped more in the circuit containing only one cell, indicating that more charge flowed in that circuit than in the other during the same time.

A hydrogen cell does not destroy charge. However, its inclusion in a circuit reduces the quantity of charge that flows through the circuit in a given time. In the example shown in Figure 10.9, the additional cell reduced the volume of hydrogen produced in 10.0 minutes from 24.8 cm^3 to 14.5 cm^3. In other words, the production of hydrogen dropped from

$$\frac{24.8 \text{ cm}^3}{10.0 \text{ min}} = 2.48 \text{ cm}^3/\text{min}$$

to

$$\frac{14.5 \text{ cm}^3}{10.0 \text{ min}} = 1.45 \text{ cm}^3/\text{min}$$

when the second hydrogen cell was added to the circuit. We conclude, therefore, that the quantity of charge that flowed through the circuit in one minute was less with two cells than it was with one cell.

Such an effect on the behavior of the circuit is a very undesirable feature of the hydrogen cell as a charge meter. In general, we want any measuring instrument to have as small an effect as possible on the system to which it is applied. If this is not so, then as a result of the measurement, we have a very different system than before.

Suppose, for example, that you wish to measure the pressure of the air inside an inflated bicycle tire. A common tire pressure gauge allows some air to escape from the tire when the pressure is measured. Consequently the pressure inside the tire is reduced. As long as the gauge

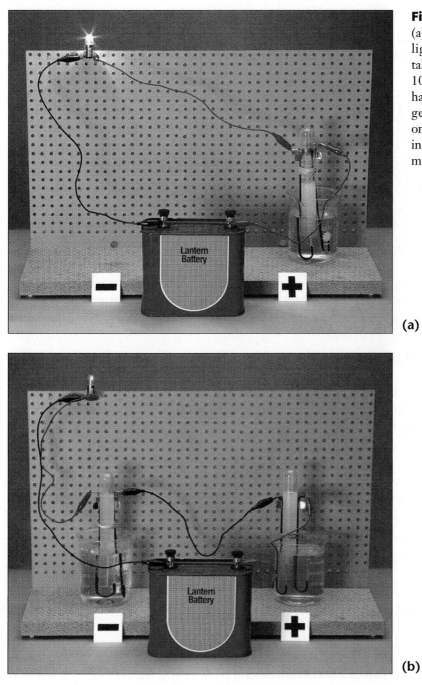

Figure 10.9
(a) One hydrogen cell in series with a light bulb. When the photograph was taken, the circuit had been operating for 10 minutes and 24.8 cm^3 of hydrogen had been produced. (b) With two hydrogen cells in series with the light bulb, only 14.5 cm^3 of hydrogen was produced in 10 minutes. Note that the light bulb is much dimmer than in (a).

(a)

(b)

removes only a small fraction of the air in the tire, you can disregard the small pressure drop caused by the use of the gauge. However, a pressure gauge that would withdraw a large fraction of the air from the tire would be useless.

Similarly, we would like to avoid the use of a charge meter that has a large effect on the charge flow. It is worthwhile, therefore, to look for other ways to measure the quantity of charge passing through a point in an electric circuit.

A convenient way to measure charge that eliminates this difficulty is to use a clock and an instrument called an *ammeter*. An ammeter measures the amount of charge that flows through a circuit per unit time. Figure 10.10 shows the circuit of Figure 10.9 with one and two ammeters instead of hydrogen cells connected in series. Notice that adding a second ammeter changed neither the brightness of the bulb nor the reading of the first ammeter. The additional ammeter had no measurable effect on the circuit.

13. **Why are the masses of thermometer bulbs always much smaller than the masses of the objects whose temperatures they are made to measure?**

14. **Suggest an experiment to determine whether adding a light bulb to a series circuit reduces the charge that flows around the circuit in a fixed time interval.**

10.7 Charge, Current, and Time

To see how you can use a clock and an ammeter, which measures the flow of charge per unit of time, to determine the total quantity of charge, first consider the following situations. A worker is paid $5 an hour or, to put it another way, she is paid at the rate of 5 dollars per hour. If she works for 8 hours, her total earnings will be

$$5 \text{ dollars/hour} \times 8 \text{ hours} = 40 \text{ dollars.}$$

Similarly, suppose 3 gallons of water flows into a pool every second and that this flow continues for 60 seconds. The amount of water that flows into the pool in 60 seconds is, therefore,

$$3 \text{ gallons per second} \times 60 \text{ seconds} = 180 \text{ gallons.}$$

These examples suggest a general relationship:

$$\text{Amount} = \text{rate} \times \text{time.}$$

We shall use this general relationship to measure the electric charge that passes through a point in a circuit:

$$\text{Amount of charge} = \text{rate of flow of charge} \times \text{time.}$$

The rate of flow of electric charge, which is what an ammeter measures, is called the electric current, so

$$\text{Charge} = \text{current} \times \text{time of flow.}$$

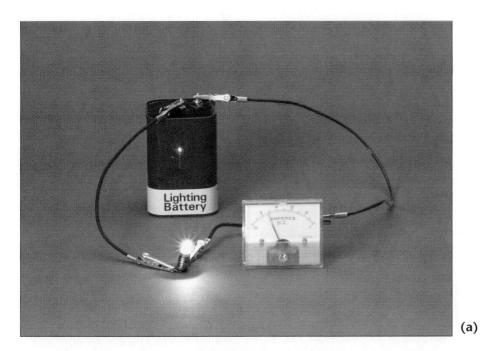

(a)

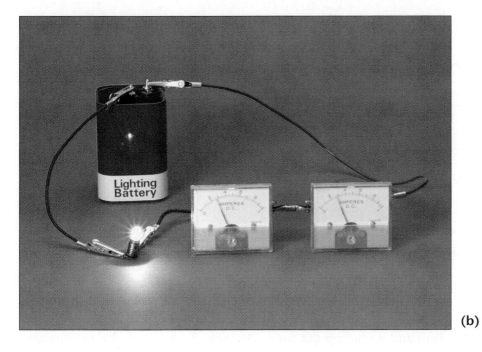

(b)

Figure 10.10

(a) One ammeter in series with a light bulb. (b) Adding another ammeter to the circuit in (a) changes neither the brightness of the bulb nor the charge per unit time (measured in amperes on the ammeter scales) through the circuit.

An ammeter measures the current in units called amperes (A). The charge when measured by the ammeter and a clock is then expressed in units of ampere-seconds, or A·s for short:

Charge (ampere-seconds) = current (amperes) × time (seconds)

Charge can be measured either with a hydrogen cell or with an ammeter and a clock. Therefore there must be a relationship between the quantity of charge measured in ampere-seconds and the quantity of charge measured in cm³ of hydrogen. This relationship is the subject of the next experiment.

†15. How many ampere-seconds of charge flow through an ammeter if it reads
   a. 2.0 amperes for 10 seconds?
   b. 0.4 amperes for 3.0 minutes?
   c. 6.0 amperes for 12.0 seconds?

16. An ammeter reads 0.75 A. How long must the current be flowing for 20 ampere-seconds of charge to flow through the meter?

17. How could you use an automobile speedometer and a clock to measure distance traveled?

EXPERIMENT
10.8 Measuring Charge with an Ammeter and a Clock

Figure 10.11 shows a hydrogen cell and an ammeter connected in series. You can use the circuit shown to find the relationship between charge measured in ampere-seconds and charge measured in cm³ of hydrogen. Measure the volume of hydrogen and the current at the end of every minute until the graduated cylinder is nearly full of hydrogen. Then you can use your data to make a graph of charge measured in ampere-seconds as a function of charge measured in cm³ of hydrogen.

Since the current may change during the experiment, you will have to know the average current. To find it, you can add all the ammeter readings up to and including the time you make a given volume reading and then divide by the number of readings. This procedure is valid as long as you have measured the current at equal time intervals.

Use your data to draw a graph of charge expressed in ampere-seconds as a function of charge expressed in cm³ of hydrogen. Compare your graph with those of your classmates.

• What volume of hydrogen is produced by a flow of 1.0 A·s of charge?

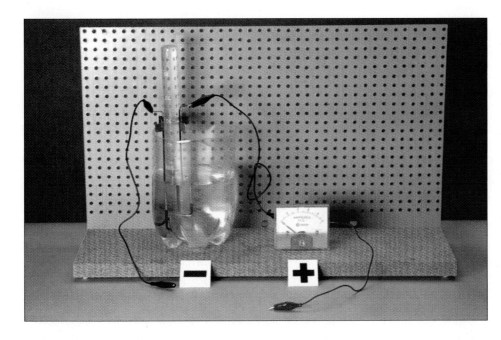

Figure 10.11
A hydrogen cell and an ammeter connected in series. A graduated cylinder is placed over the negative electrode to collect the hydrogen so that the volume of gas can be measured directly. When you connect this circuit, be sure that the positive terminal of the ammeter is connected to the positive terminal of the battery or power supply.

- Does the volume of hydrogen produced by a flow of 1.0 A·s of charge depend on the number of flashlight cells or the setting of the power supply used in the circuit?

- Does it depend on the current?

The volume of hydrogen produced by 1.0 ampere-second of charge can vary. Both the pressure and the temperature of the gas will affect the result. For example, you get quite different results if you do the experiment in Denver, Colorado, on a hot day and in Boston, Massachusetts, on a cold day. However, the mass of hydrogen produced in each case during 1 second is the same in both places. When one ampere flows through the cell, 1.04×10^{-5} g of hydrogen is produced per second, regardless of the temperature or pressure of the gas.

18. Suppose the ammeter in the circuit of Figure 10.11 measured charge instead of charge per unit time. How would the position of the needle be affected as hydrogen was produced?

19. In doing Experiment 10.8, group I measured the current at the end of every minute during a 5-minute run and recorded the data in the table.

Time (min)	Current (A)
1	0.50
2	0.80
3	0.81
4	0.82
5	0.82

Group II took only two readings of current during the 5-minute run and did not pay any attention to the time of the readings. The group recorded the following data:

First reading   0.50 A
Second reading   0.82 A

a. What was the charge through the circuit during each interval of 1 minute in the experiment of group I?

b. What was the total charge through the circuit in the 5-minute run in the experiment of group I?

c. What was the average current during the 5-minute run in the experiment of group I?

d. Can group II answer any of these questions for its experiment?

20. From the data of Experiment 10.8, calculate how many cubic centimeters of hydrogen at room temperature you would collect if you ran a hydrogen cell for 5 minutes with an ammeter reading of 0.5 A.

FOR REVIEW, APPLICATIONS, AND EXTENSIONS

21. Examine a light bulb at home to see if you can find the two contacts that make the bulb operate.

22. There are no connecting wires in flashlights. Explain how it is possible for any charge to flow.

23. Suppose you have three identical light bulbs, some connecting wire, and a battery. Make a sketch of all the different possible ways to connect the battery and bulbs, using all three bulbs each time. Label which circuits are series circuits and which are parallel circuits.

24. Draw a circuit diagram that includes some identical hydrogen cells. Show how you could collect exactly three times as much hydrogen in one cell as in one of the others during the same time.

25. "Electric charge is neither created nor destroyed when it flows around a circuit." What happens to the electric charge when the circuit is disconnected at one point?

26. Someone suggests that the brightness of a bulb depends on the total quantity of charge that passes through it, and not on the quantity of charge per unit time. How would you disprove this?

27. What would happen if the connections to the battery were interchanged (reversed) in the circuit shown in Figure 10.11?

28. In a laboratory experiment, a student measures the current in a circuit as a function of time. What average current flowed through the circuit if Figure D is the graph of current versus time? If Figure E is the graph of current versus time?

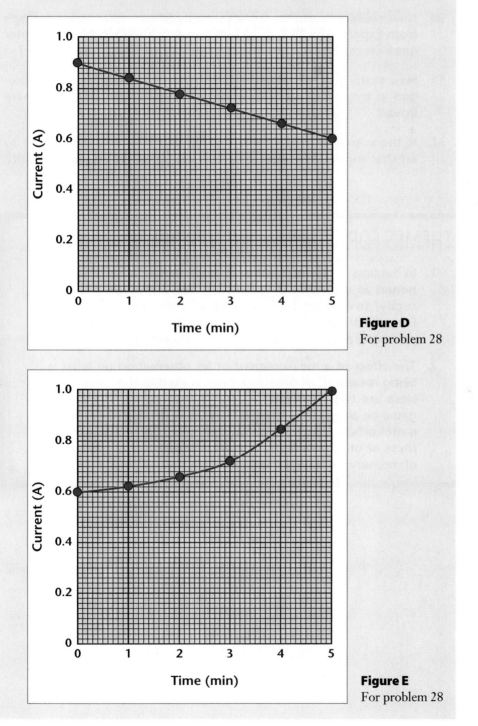

Figure D
For problem 28

Figure E
For problem 28

29. A hydrogen cell and two ammeters are connected in series. One ammeter reads 0.50 A and the other reads 0.65 A. In 10 minutes 52 cm³ of hydrogen is produced. Using your graph from Experiment 10.8, decide which ammeter should be set aside for repair.

30. How could you use an ammeter, a hydrogen cell, and the graph from Experiment 10.8 as a clock for which each cubic-centimeter mark on the 50-cm³ graduated cylinder represents one minute?

31. How many seconds would it take to produce one gram of hydrogen in a hydrogen cell through which a current of one ampere flows?

32. Is there any evidence from the experiments you have done so far that suggests in what direction charge flows around a circuit?

THEMES FOR SHORT ESSAYS

1. In Section 10.1 electric charge and temperature are mentioned as examples of quantities that are measured indirectly. Invent an indirect way of measuring appetite. Describe the procedure in detail. Discuss the assumptions you make and the reliability of the procedure.

2. The effect of a measurement or an observation on what is being measured or observed is not limited to the laboratory. Here are two examples. How can the score of a football game be affected by the watching crowd? Does a highway-patrol officer affect the number of speeding motorists? Use these or other examples to write a brief essay on the effect of measurement on what is being measured.

Chapter 11

Atoms and Electric Charge

11.1 The Charge per Atom of Hydrogen and Oxygen

In the last chapter we used the amount of hydrogen produced by electrolysis as a measure for the quantity of electric charge that flows in a circuit. The greater the amount of charge that passed through the cell, the greater the amount of hydrogen and oxygen produced. This indicates a connection between matter and electric charge.

To investigate this connection further, we shall study electrolytic cells where the passage of electric charge results in the production of other elements on an electrode. Since we believe these elements accumulate on an electrode atom by atom, it will be useful to compare the quantities of charge needed to release single atoms of various elements.

In the case of hydrogen, a flow of 1 ampere-second of charge releases 1.04×10^{-5} g of the gas. From Section 9.7 you know that the mass of a hydrogen atom is 1.66×10^{-24} g. Thus, the number of hydrogen atoms produced by a flow of 1 A·s is equal to the mass of hydrogen produced divided by the mass of a single atom:

$$\text{Number of atoms} = \frac{\text{mass of sample}}{\text{mass of single atom}}$$

$$= \frac{1.04 \times 10^{-5} \text{ g}}{1.66 \times 10^{-24} \text{ g}} = 6.3 \times 10^{18} \text{ atoms.}$$

If 1 A·s of charge releases 6.3×10^{18} atoms of hydrogen, then the charge required to release one atom of hydrogen will be

$$\frac{1.0 \text{ A·s}}{6.3 \times 10^{18} \text{ atoms}} = 1.6 \times 10^{-19} \text{ A·s.}$$

How much charge is needed to produce one atom of oxygen when water is electrolyzed?

From the experiment on the decomposition of water (Section 6.2), you calculated that the mass ratio of hydrogen to oxygen in water is 1 to 8. From the table of atomic masses (Table 9.2), you saw that the ratio of the atomic mass of hydrogen to that of oxygen is 1 to 16. Therefore, in a compound that has one atom of hydrogen for every atom of oxygen, the mass ratio of hydrogen to oxygen would be 1 to 16. In water this mass ratio is 1 to 8 or 2 to 16. We conclude, therefore, that in water there are two hydrogen atoms for every oxygen atom. (This yields the well-known simplest formula for water: H_2O.) In other words, if we decompose any amount of water, we release twice as many atoms of hydrogen as of oxygen.

Since the same quantity of electric charge passes through both electrodes, each atom of oxygen must require twice the charge needed to release one atom of hydrogen. Therefore we conclude that it takes

$$2.0 \times (1.6 \times 10^{-19})\ \text{A·s}$$

to release one atom of oxygen.

†1. During the electrolysis of water, which collecting tube contains the greater number of atoms of gas at a given time?

2. A charge of 1 A·s yields 1.04×10^{-5} g of hydrogen. How many grams of oxygen will 1 A·s of charge release? What volume will the oxygen occupy?

EXPERIMENT
11.2 The Electroplating of Zinc

We have just seen that there is a simple 2-to-1 ratio for the quantities of charge needed to release one atom of oxygen and one atom of hydrogen. How does the charge per atom compare for different elements?

In this experiment you will determine the quantity of charge needed to release one atom of zinc from a solution containing zinc. Since zinc is a solid, it will be deposited as a solid on the electrode. To determine the mass released, you can mass this electrode before and after you electrolyze the solution. Then you can find the charge per atom.

$$\text{Charge per atom} = \frac{\text{charge needed to release sample of element}}{\text{number of atoms in sample}}$$

With an ammeter and a clock, you can measure the charge in ampere-seconds. From the change in mass of the electrode and the mass of a zinc atom, you can determine how many atoms of zinc were plated during the experiment. The mass of a zinc atom is given in Table 9.2.

The zinc-plating cell consists of two zinc electrodes and a solution containing zinc. After massing each of the two zinc electrodes, you can connect the zinc-plating cell to an ammeter as shown in Figure 11.1 on page 216. Do not connect the battery to the ammeter and the plating cell until you are ready to time the run.

Begin by connecting the circuit to only one flashlight cell. Quickly increase the number of cells until the current is at a maximum with the ammeter needle on the scale.

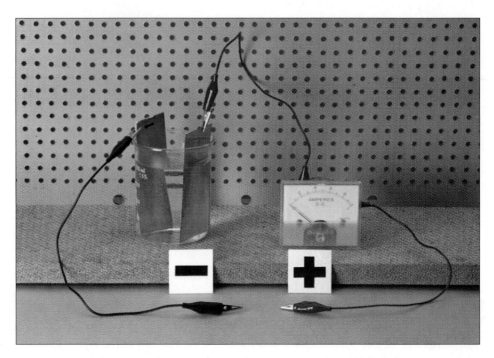

Figure 11.1
A zinc-plating cell and an ammeter connected in series. The wire lead on the right comes from the ammeter terminal marked "+" and is connected to the positive terminal of the battery (not shown in the photograph).

A run of 20 to 25 minutes will deposit enough zinc to be massed on the balance. Since the current may change during the run, a current reading every minute will be useful.

When the run is over, rinse both electrodes by dipping them in a beaker of water. Gently dry the electrodes with paper towels and mass each one on your balance. Compare the changes in mass for the two electrodes. Use Table 9.2 on page 189 to find how many atoms of zinc were deposited.

From the average current that flowed through the circuit and the duration of the run, you can calculate the electric charge that was used to plate out the zinc.

- What do you find for the value of the charge, in ampere-seconds?

- How much charge was needed to deposit one atom of zinc?

- How does this compare with the charge needed to release one atom of hydrogen?

3. A nickel-plating cell was run for 10 min at an average current of 0.80 A. It was found that 0.15 g of nickel was deposited on one electrode.
   a. What is the mass of one atom of nickel?
   b. How many nickel atoms were deposited?
   c. What charge flowed, in ampere-seconds?
   d. What is the charge per atom for nickel in this experiment?

4. In Experiment 11.2, The Electroplating of Zinc, one student calculated the charge in ampere-seconds for each minute, added all the results, and then divided by the total time in minutes. Another student found the average current and multiplied by the total time in seconds. Which method is correct for determining the total charge?

5. Suppose a nickel-plating cell is run for 5 min at a current of 0.6 A. Then the battery leads are reversed, and the cell runs for 5 min more at the same current. What happens at the electrodes? Would the result be the same for a cell electrolyzing water?

11.3 The Elementary Charge

Hydrogen, oxygen, and zinc are not the only elements that can be collected at the electrodes of an electrolytic cell. Other elements can also be collected this way, and the charge needed to deposit one atom can be determined. However, the procedure may in some cases be considerably more difficult than the ones you have used. For example, sodium cannot be plated out from a solution of sodium chloride in water. Hydrogen will evolve at the negative electrode, and the sodium will remain in the solution. Chlorine will bubble up at the other electrode. However, since chlorine is quite soluble in water, the amount of chlorine gas collected in the test tube is much less than the amount of chlorine actually produced.

To obtain reliable data from which to calculate the charge per atom for sodium and chlorine, something else must be done. We have to pass a current through molten sodium chloride and prevent any contact of the sodium with water vapor or oxygen. Potassium requires a similar treatment.

Despite the technical complications, many elements have been produced by electrolysis, and the charge required per atom has been determined. Table 11.1 on page 218 shows the ratios of the charge per atom for some common elements to the charge per atom of hydrogen, as obtained from experiments more precise than those you have done.

Table 11.1 strongly suggests that the electric charge needed to produce one hydrogen atom by electrolysis has a fundamental significance. First of all, it is the smallest quantity of charge that is involved in the electrolysis of any material. But, more than that, the charge required to deposit one atom of any element is equal to the charge per atom of hydrogen or to a small whole number times this charge. There is no element that requires, say, 2.5 times this charge to release one atom. For these reasons the charge needed to release one hydrogen atom, 1.60×10^{-19} A·s, is known as the *elementary charge*. It is the smallest quantity of charge now known to occur in nature.

Table 11.1

Element	Symbol	Charge per atom of element / Charge per hydrogen atom
Aluminum	Al	3.00
Bromine	Br	1.00
Calcium	Ca	2.00
Chlorine	Cl	1.00
Chromium	Cr	6.00
Iodine	I	1.00
Lithium	Li	1.00
Magnesium	Mg	2.00
Mercury	Hg	2.00
Nickel	Ni	2.00
Oxygen	O	2.00
Potassium	K	1.00
Silver	Ag	1.00
Sodium	Na	1.00

About 170 years ago Michael Faraday in England carried out electrolysis experiments similar to the ones you have done. He found that the mass of an element deposited on an electrode is proportional to the quantity of charge that flows through the circuit. But Faraday, who was not convinced that matter was composed of atoms, did not suggest an elementary unit of charge.

The first to propose the idea of an elementary charge was Hermann Helmholtz in Germany. In 1881 he related the results of Faraday's experiments to the atomic theory with this bold declaration: "If we accept the hypothesis that the elementary substances are composed of atoms we cannot avoid the conclusion that electricity also . . . is divided into definite elementary portions, which behave like atoms of electricity."

So far, we have seen that the units we use to measure physical quantities are quite arbitrary: the centimeter is as good a unit of length as the inch. Nowhere have we seen any preference in nature for one unit over another. The electric charge is different. Nature provides us with its own fundamental unit, the elementary charge.

6. A hydrogen cell is connected in series with an aluminum-plating cell.
   a. How many atoms of hydrogen will be produced for every atom of aluminum deposited?
   b. How many grams of hydrogen will be produced for every gram of aluminum deposited?

7. Will an electric current which passes through a zinc-plating and a magnesium-plating cell in series ever deposit the same mass of metal in each cell?

8. What is the significance of recording the ratios in Table 11.1 as 3.00, 1.00, 2.00, . . . , and not simply as 3, 1, 2, . . . ?

11.4 The Elementary Charge and the Law of Constant Proportions

Consider two elements A and B that combine to form a compound. If A and B require equal numbers of elementary charges per atom in electrolysis, then they combine according to the formula AB, that is, one atom of A for every atom of B. On the basis of this generalization, we would predict from Table 11.1 that hydrogen and iodine combine according to the formula HI, and that mercury and oxygen form the oxide HgO. When these compounds are analyzed, this is indeed found to be their composition.

Suppose element A requires one elementary charge per atom and element B requires two. If they combine at all they will do so according to the formula A_2B or (BA_2), since each time two elementary charges release one atom of B, two atoms of A are released. The compounds H_2O, Na_2O, and $MgCl_2$ are examples of this kind of composition.

Accurate measurements of the mass ratios of the elements in these and many other compounds confirm predictions based on Table 11.1. Additional examples are given in Table 11.2.

Table 11.2

Element	Elementary charge per atom	Element	Elementary charge per atom	Simplest formula of compound
Iron	2	Chlorine	1	$FeCl_2$
Silver	1	Chlorine	1	AgCl
Hydrogen	1	Bromine	1	HBr
Lithium	1	Oxygen	2	Li_2O
Silver	1	Oxygen	2	Ag_2O

To sum up, once we have found the number of elementary charges per atom for various elements by electrolysis, we know how many atoms of one element would combine with one atom of another element. This is true even for compounds that cannot be electrolyzed. The number of elementary charges per atom and the atomic mass fix the proportions in the law of constant proportions. When we first encountered this law, we had no way of relating mass ratios to anything else. Now we see that they are closely connected to atomic masses and the number of elementary charges per atom.

But elements form many compounds. Does this mean that a given element always has the same charge per atom in all its compounds? We shall try to answer this question in the next experiment.

9. To write the simplest formula for water you can use either the data on the combining masses and atomic masses of hydrogen and oxygen, or the data on the charge per atom for these two elements in electrolysis. Upon what conservation laws are these arguments based?

10. Using Table 11.1, write the simplest formula for each compound.
    a. magnesium oxide
    b. chromium oxide
    c. aluminum oxide

†11. Suppose element A combines with element B according to the simplest formula A_2B. Element B has been found to require two elementary charges per atom in electrolysis. If element A could be plated out in electrolysis, how many elementary charges per atom would it require?

12. In Experiment 6.4, The Synthesis of Zinc Chloride, you showed that the ratio of zinc reacted to product formed was 0.48. Using the data from Tables 9.2 and 11.1, what ratio do you obtain?

EXPERIMENT
11.5 Two Compounds of Copper

The two compounds of copper to be investigated are copper sulfate (a blue solution) and copper chloride (an almost colorless solution).

Mass two copper electrodes after marking them with identifying letters and your initials.

Connect two copper-plating cells, each with copper electrodes, in series with an ammeter as shown in Figure 11.2. In each cell the negative electrode is the one you have marked for identification. Add the blue solution to one cell and the nearly colorless solution to the other.

Using the same procedure as in Experiment 11.2, adjust the current close to 1 A in order to deposit in 15 minutes enough copper to mass. Be sure you record the current every minute during the run.

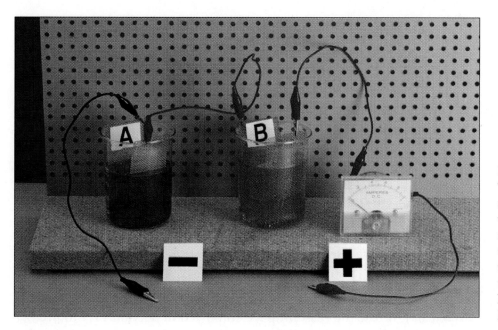

Figure 11.2
Two copper-plating cells connected in series. The wire lead on the right runs from the ammeter terminal marked "+" to the positive terminal of the battery (not shown). Note which electrodes are labeled for identification.

After carefully rinsing and drying the marked electrodes, mass them and find the gain in mass of each.

- How many atoms of copper were deposited in each electrolytic cell by the charge that flowed through the circuit?

- Without further calculation, what can you say about the quantity of charge required to plate out one atom of copper from the blue solution, compared with the charge needed to plate out one atom of copper from the colorless solution?

- From the average current that flowed through the circuit and the duration of the run, what is the number of elementary charges needed to plate out one atom of copper from each of the two solutions?

There are two oxides of copper—that is, two compounds containing only copper and oxygen.

- On the basis of the results of this experiment, what do you predict are their simplest formulas?

†13. Suppose the electrodes in one of the cells in the experiment described above were bent so that they were touching throughout the run. What effect would this have on the results?

11.6 The Law of Multiple Proportions

There are other elements besides copper that require different numbers of elementary charges per atom to release them, depending on the compound from which they are plated out. Table 11.3 lists some of them, including a few that appear in Table 11.1.

Table 11.3

Element	Symbol	Number of elementary charges per atom for different compounds
Chromium	Cr	2, 3, 6
Iron	Fe	2, 3
Mercury	Hg	1, 2
Tin	Sn	2, 4

We can now use the reasoning we employed in Section 11.4 to predict the simplest formulas for different compounds containing these elements. Some examples are shown in Table 11.4.

Table 11.4

Element	Elementary charge per atom	Element	Elementary charge per atom	Simplest formula of compound
Iron	2	Chlorine	1	$FeCl_2$
Iron	3	Chlorine	1	$FeCl_3$
Tin	2	Chlorine	1	$SnCl_2$
Tin	4	Chlorine	1	$SnCl_4$
Chromium	2	Oxygen	2	CrO
Chromium	3	Oxygen	2	Cr_2O_3
Chromium	6	Oxygen	2	CrO_3

Consider the two chlorides of iron in Table 11.4. What is the ratio of the mass of iron in $FeCl_2$ to the mass of iron in $FeCl_3$ for the same mass of chlorine?

Suppose that a cell containing a solution of $FeCl_2$ and a cell containing a solution of $FeCl_3$ are connected in series. Suppose further that all the chlorine and all the iron produced were collected while a current passed through the cells. Since in both chlorides the chlorine carries one elementary charge per atom, the same mass of chlorine would be collected in both cells. However, in $FeCl_3$, 3 elementary charges were needed to plate out an atom of iron, while only 2 elementary charges were needed in $FeCl_2$. Therefore, the mass of iron plated out from $FeCl_2$ is $\frac{3}{2}$ as large as the mass of iron plated out from $FeCl_3$ for the same mass of chlorine. This is a ratio of two small whole numbers: 3 and 2.

Table 11.4 predicts the simplest formulas for two chlorides of tin. If this prediction is correct, then the mass of chlorine in $SnCl_4$ that combines with a given mass of tin is twice the mass of chlorine that combines with the same mass of tin in $SnCl_2$. Measurements show that the predicted mass ratios are correct.

In general, for any two elements that form more than one compound, the ratio of the masses of one element that combine with a fixed mass of the other element is a ratio of small whole numbers. This generalization is known as the *law of multiple proportions*.

You have now studied two laws concerning the formation of compounds: the law of constant proportions (Section 6.5) and the law of multiple proportions. The first law applies to a single compound of any two given elements. To describe the formation of more than one compound by the same two elements, we need the second law. Thus the law of multiple proportions is an extension of the law of constant proportions.

When two elements form more than one compound, it is often difficult to synthesize one of the compounds without forming some of the others at the same time. The relative amounts of the various compounds depend on the relative quantities of the elements, the temperature, and other conditions. In such cases, experiments will yield different ratios for the elements in what may be mistaken for a single compound.

Part of the difficulty of originally establishing the law of constant proportions was the formation of more than one compound by the same two elements. This is one of the reasons why the combining of copper and sulfur seemed to violate the law of constant proportions. (See Section 6.5.) Another complication occurs when the sulfur is present in excess. The compounds formed dissolve in the excess sulfur to form "frozen" solutions. This may give the material the appearance of a uniform solid, which is easily mistaken for a single compound. It is very difficult to separate the newly formed components from a "frozen" solution.

14. There are two compounds containing only chlorine and mercury. What would you predict as the simplest formulas for these two compounds?

15. Using the data in Table 11.3, write the simplest formula for the two oxides of mercury.

16. Which of the following simplest formulas do you think describe real compounds? (See Tables 11.1 and 11.3.)
    a. SnH_3
    b. $NaCl_2$
    c. SnO_2

FOR REVIEW, APPLICATIONS, AND EXTENSIONS

17. How much charge must pass through a hydrogen cell to produce a test tube of hydrogen (about 35 cm³)? Use the class results for Experiment 10.8, Measuring Charge with an Ammeter and a Clock.

18. In the electrolysis of water we concluded that water must be continually added to the cell to maintain the same water level. From your observations of metal-plating cells, do you think that the solutions involved must be replenished to maintain the same liquid level?

19. A student ran a lead-plating experiment with a battery that was known to contain cells that were weak from constant use. How would this affect the calculations of charge per atom for lead?

20. In a metal-plating experiment, the following measurements and calculations were obtained.
    A Mass of metal deposited
    B Number of atoms deposited
    C Average current
    D Number of charges that flowed
    E Charge per atom of metal
    How would each of the above observations be affected if
    a. the length of time for the run was doubled?
    b. the electrodes were moved closer together?
    c. an additional cell was used in the battery?
    d. the electrodes were not rinsed before the final massing?

21. Suppose three electrolytic cells are connected in series. Hydrogen is collected in a test tube, and aluminum and nickel are plated out. If 1 g of hydrogen is collected, what would be the mass of the other elements plated out?

22. The decomposition of uranium chloride yields UCl_4 as the simplest formula for the compound. How many elementary charges per atom do you expect to find in plating uranium from a solution?

23. Some of the first precise measurements of the atomic masses of elements were made with plating cells. Suppose a gold-plating cell and a silver-plating cell are connected in series to a battery. When 1.00 g of silver is deposited, 1.83 g of gold is plated out in the gold cell. The mass of a silver atom is 108 u and one elementary charge is needed to deposit one silver atom. Determine the mass of a gold atom. What important assumption did you make? How does your value for the mass of a gold atom compare with that in Table 9.2?

24. Suppose element A combines with element B according to the simplest formula A_2B. Element B combines with element C in the simplest formula CB. What might be the simplest formula for a compound made from A and C?

25. A hydrogen-oxygen cell and a copper-plating cell that contains the blue solution copper sulfate are connected in series until 10 cm^3 of hydrogen is collected. Describe the steps you would take to calculate how much copper was plated during this time.

26. A mass of 16 g of oxygen combines with 63.5 g of copper to form CuO.
    a. What is the ratio of the mass of copper to the mass of oxygen?
    b. What is the ratio of the mass of copper to the mass of oxygen in Cu_2O?

27. Carbon can combine with chlorine to form three different compounds, CCl_4, C_2Cl_4, and C_2Cl_6. The ratio of the mass of carbon to the mass of chlorine in CCl_4 is 0.0845. What is the ratio of the mass of carbon to that of chlorine in the other two compounds?

28. Four students suggested the following pairs of formulas for the two chlorides of copper:

	Brown chloride	Green chloride
Student 1	$CuCl_2$	$CuCl$
Student 2	$CuCl$	Cu_2Cl
Student 3	Cu_3Cl_2	$CuCl$
Student 4	Cu_3Cl_4	Cu_3Cl_2

    a. Which pair or pairs are possible formulas for these compounds?
    b. Can you decide which pair gives the correct formulas for the chlorides of copper? Why or why not?

THEME FOR A SHORT ESSAY

Suppose that a dish, claimed to be made of solid silver, might only be silver-plated. However, the density of the material is very close to that of silver. Describe an experiment that will determine, with minimal damage to the dish, whether the dish is silver-plated or solid silver.

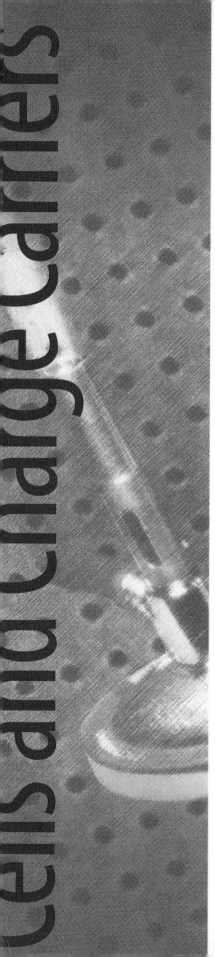

Chapter 12
Cells and Charge Carriers

From our experiments with electrolytic cells, we have drawn some far-reaching conclusions about the flow of charge in a solution. In these experiments a battery of flashlight cells caused an electric charge to flow through various solutions. What happens inside a flashlight cell when we draw current from it? It is not easy to answer that question from simple observation. However, a similar process takes place inside another kind of cell, called a Daniell cell. A Daniell cell can be examined in detail.

EXPERIMENT
12.1 The Daniell Cell

A Daniell cell is constructed as shown in Figure 12.1. Two solutions are placed in the same container, but they are separated by a thin wall of parchment paper through which they can mix only very slowly. Notice that this cell differs from the plating cells you have used. Not only are two solutions used, but the two electrodes are made of different metals. A photograph of the cell is shown in Figure 12.2.

When a Daniell cell is providing current, is there a change in mass of the two electrodes? If so, is the charge per atom needed to dissolve or plate material the same as when zinc and copper are plated out in electrolytic cells? You will answer these questions in this experiment.

When you have made a cell like that in Figure 12.1, mass the electrodes and put them in place in the cell. In order to minimize mixing of the solutions, do not pour the two solutions into the cell until you are ready to record the current and time.

When the cell has been filled, connect it to the ammeter. Adjust the separation of the electrodes until the current is about 0.8 A. You can do this by rotating the clamps. It is best to record the current every minute for 15 minutes.

After rinsing both electrodes, dry and mass them.

• What is the change in mass of each electrode?

• What charge in ampere-seconds flowed through the circuit?

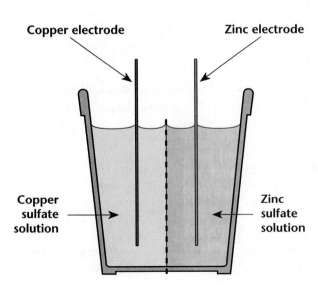

Copper electrode

Zinc electrode

Copper sulfate solution

Zinc sulfate solution

Figure 12.1

A diagram showing the construction of a simple Daniell cell. The solutions are contained in a Styrofoam cup that is separated into two compartments by a parchment paper partition, shown as a dashed line.

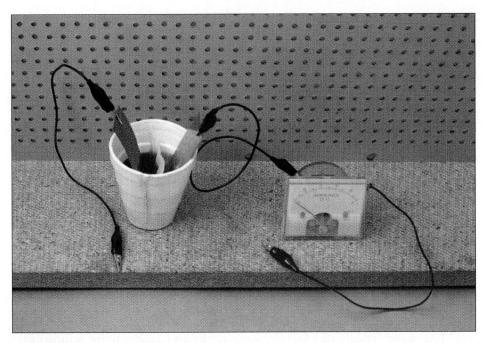

Figure 12.2
A Daniell cell with one terminal connected to an ammeter. The cell will begin operating when the circuit is completed by clipping together the two alligator clips lying in front of the meter. The left half of the cell contains the copper electrode and the dark-blue copper sulfate solution. The right half contains the zinc electrode and the colorless zinc sulfate solution. The copper electrode is connected to the "+" terminal of the ammeter.

- How many atoms of zinc dissolved and how many atoms of copper plated out?
- What charge, in ampere◊seconds, was needed to dissolve one atom of zinc and plate out one atom of copper?
- Restate your answers to the preceding question in elementary charges per atom. What do you find?
- How do these values compare with those you found when these metals were plated out in electrolytic cells in the experiments in Chapter 11?

1. A Daniell cell will deliver current for a long time but will eventually stop. What do you think could happen to the electrodes or solutions to cause the cell to stop?

2. Suppose a Daniell-cell experiment gives 1.7 elementary charges per atom for both zinc and copper. What would be the most likely source of this error?

EXPERIMENT

12.2 Zinc and Copper in Different Solutions

The Daniell cell you just used contained two different solutions as well as two different electrodes. You may have wondered why it was necessary to use two solutions and keep them separated by a special partition. You can answer this question by observing the effect of placing each of the electrodes used in the Daniell cell in each of the two solutions. Be sure to rinse each electrode in water each time you remove it from a solution.

First try dipping the copper electrode into copper sulfate solution.

- What happens?

Now try the copper electrode in zinc sulfate solution.

- Does anything happen?

- What is the result of placing the zinc in the zinc sulfate solution and in the copper sulfate solution?

- In which of these cases, using just a single electrode, did you observe an effect?

Now connect two zinc electrodes to an ammeter and then dip them briefly first into one solution and then into the other.

- What do you observe? If the ammeter needle moves in the wrong direction, reverse the connections.

Try two copper electrodes, connected by an ammeter, in each of the solutions.

- What happens?

Now connect a zinc electrode and a copper electrode to an ammeter and dip the two electrodes into each of the two solutions in turn.

- In which of these cases, using two electrodes and an ammeter, did you observe an effect?

Suppose you have a cell consisting of electrodes of copper and zinc in a copper sulfate solution.

- What will happen in this cell when it is not being used?

- Could you use such a cell to determine the number of elementary charges per atom when zinc dissolves and copper plates out?

- From the observations you made in this experiment, why are the two solutions separated by parchment paper in the Daniell cell?

12.3 Flashlight Cells

The flashlight cells that you have used in your experiments, though quite complex, have much in common with a Daniell cell or a simple zinc, copper, and copper sulfate cell. Like all current-generating cells, a flashlight cell consists of two different electrodes and a water solution of a compound.

Figure 12.3 shows the construction of a flashlight cell. The positive electrode is a cylindrical carbon rod in the center of the cell. Surrounding the electrode is a paste that consists primarily of a water solution of ammonium chloride mixed with powdered carbon and manganese dioxide. The negative electrode is a cylindrical zinc can that serves not only as an electrode but also as a container for the carbon electrode and the other materials. This kind of cell is often called a *dry cell*, because the

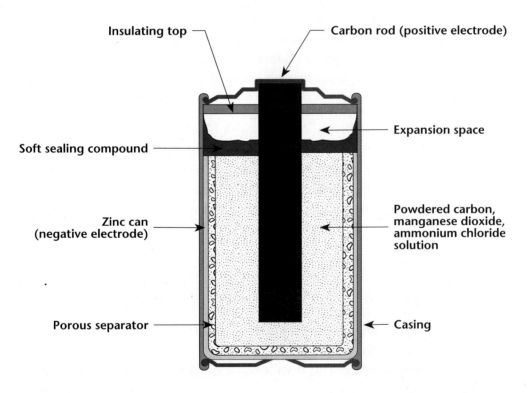

Insulating top

Carbon rod (positive electrode)

Expansion space

Soft sealing compound

Powdered carbon, manganese dioxide, ammonium chloride solution

Zinc can (negative electrode)

Porous separator

Casing

Figure 12.3

A cross section of a flashlight cell. The porous separator keeps the carbon and manganese dioxide from coming into contact with the zinc but allows ammonium chloride solution to pass freely. The zinc can is enclosed in a tight-fitting tube that holds the insulating top in place and insulates the negative zinc can from adjacent cells or adjacent metal objects such as the case of a flashlight.

solution, held in the porous paste and sealed in the zinc can, will not leak out of the cell.

When charge flows in this type of cell, zinc dissolves, forming zinc chloride. Manganese dioxide then reacts with the solution to form another oxide of manganese.

If the current supplied by a dry cell is too large, hydrogen gas accumulates around the carbon electrode. As this gas accumulates, it blocks the flow of charge and the cell runs down rapidly, to the point where it cannot supply enough current to be useful. However, such a rundown cell partially recovers in a few hours when it is disconnected from a circuit, because the hydrogen gas slowly leaks out of the cell.

After a dry cell has been used for some time, insoluble, nonconducting compounds form that hinder the flow of charge through the cell. Then it cannot supply enough current to be useful and must be discarded. A newly manufactured cell, even when not used, eventually becomes useless. Water evaporates from the cell and reactions go on, though very slowly, in the cell even when it is not connected to a circuit.

There are many other types of cells that are used to move charge through a circuit. Some of these types of cells can be recharged. In recharging, the reactions that go on when the cell is used are reversed. To reverse the reactions a charge is sent through the cell in the direction opposite to that in which the cell moves charge through a circuit. This restores the electrodes and solution close to their original condition. The cell can then be used again to operate a circuit.

All cells, regardless of the materials of which they are made, are basically the same. They have two dissimilar electrodes in solutions, and when they provide current, some compounds are decomposed and new ones are formed.

3. Suppose a current is drawn from a flashlight cell.
   a. How much charge in ampere-seconds flows if the cell delivers 1 A for an hour?
   b. How many elementary charges is this?
   c. How many atoms of zinc dissolve?
   d. What mass of zinc dissolves? (Use Table 9.2)

4. Suppose the circuit shown in Figure 10.9(a) is allowed to operate until 10 cm³ of hydrogen is collected.
   a. What is the mass of this quantity of hydrogen?
   b. How much zinc dissolved in the flashlight cell at the far left?
   c. How much zinc dissolved in each of the other cells?

12.4 Unintentional Cells and Corrosion

It takes careful design to produce an efficient dry cell or an automobile-battery cell that will give reasonably large currents over a long time. However, it is just as difficult to avoid creating cells that will produce very small currents. A large variety of elements and compounds will act as electrodes, and there are many solutions that will complete the cell. For example, an aluminum nail and a brass paper clip stuck into an apple will produce a current, though a small one (see Figure 12.4). Even tap water contains enough dissolved material to make a current-producing cell when two dissimilar metals are put into it. The sweat from your hands, or even saliva, will also serve as a solution for a cell.

Often, two dissimilar metals and a solution may accidentally form an unwanted current-producing cell. Such a cell results because of the tendency of some metals to react when they are in contact with other metals (or in some cases compounds) in a moist environment. For example, when a piece of iron and a piece of copper come into contact on moist earth, the iron will rust away rapidly. A sensitive ammeter would show that charge flows between the two metals where they are in contact. Unintentional cells like this are the cause of an enormous amount of corrosion damage to metal structures.

Very pure iron rusts slowly. But most iron and steel contain impurities such as carbon, some in the form of little islands embedded in the

Figure 12.4
A current-producing cell made from an apple. The two electrodes are an aluminum nail and a brass paper clip. The solution in the cell is the juice of the apple. The current produced by these cells is very small, and the scale on the ammeter is marked in millionths of an ampere.

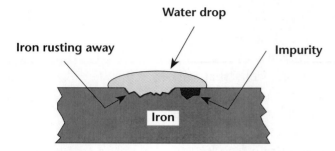

Figure 12.5
The rusting of iron is speeded by the action of small, current-producing cells created by bits of impurity embedded in its surface. In the drawing, a water drop containing a small amount of dissolved material serves as the "solution" of the tiny cell. Charge flows directly from the impurity into the iron.

surface of the metal (Figure 12.5). When the surface of the iron or steel becomes damp, many very small cells are set up. Then one of the substances—usually the iron itself—begins to dissolve.

The corrosion of iron and steel structures that are buried underground or are underwater can be considerably reduced by setting up a current-producing cell with the iron as one electrode. The other electrode can be a metal such as zinc or magnesium. When the two metals are connected by a wire, hydrogen is produced at the iron electrode, and the corrosion of the iron is stopped. The iron does not dissolve or corrode; the magnesium or zinc electrode is the one that dissolves.

Figure 12.6 shows how such a cell can be used to protect a buried iron tank from corrosion. The "solution" is the water in the soil, which always contains some dissolved substances that allow the cell to operate. The magnesium is sacrificed to protect the more valuable iron tank (Figure 12.7).

Similarly, a zinc or magnesium rod can be attached to the steel hull of a seagoing ship below the water level. The ship and the rod form a cell in which the zinc or magnesium dissolves, protecting the steel ship from corrosion.

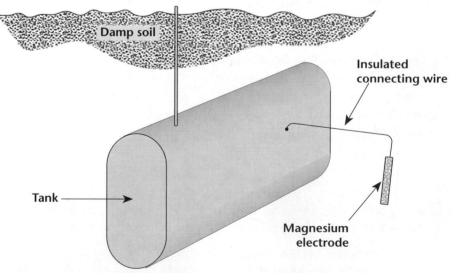

Figure 12.6
An iron tank buried in damp soil can be protected against corrosion by making it one electrode of a current-producing cell whose other electrode is a rod of magnesium. The magnesium slowly dissolves, but the iron of the tank does not. In practice, several magnesium electrodes are buried around the tank, each being connected to it by a wire.

Figure 12.7
A new magnesium electrode (right), and a badly corroded
magnesium electrode (left), which has been protecting an iron
fuel tank for many years.

Iron or steel pipe, pails, and many iron or steel sheet-metal products
are coated with a very thin layer of relatively noncorrosive zinc. Iron
coated this way is called *galvanized iron* and is much less subject to corrosion than iron alone. If the underlying iron is exposed to the atmosphere
as a result of scratches in the zinc coating, a current-producing zinc-iron
cell is formed. In this case the zinc dissolves slowly but the iron does not.
The iron is therefore protected from corrosion until enough zinc has
been consumed to expose large patches of bare iron. Zinc is more costly
than iron or steel, but only a small amount of zinc is needed to give iron
a thin protective coating.

5. A person built an aluminum canoe and used copper rivets to join
    the parts to give the canoe a more attractive appearance. What
    do you think would happen if this canoe were used in salt water?

6. Steel pails are often coated with zinc to prevent corrosion.
    What would happen if you used such a pail to store a solution
    of copper sulfate?

7. Suppose a piece of iron is part of an unintentional cell that deliv-
    ers a current of 10^{-3} A. Estimate how much iron will corrode in
    one year.

12.5 The Motion of Electric Charge Through a Vacuum

So far we have been concerned with the flow of charge in electrolytic cells and in current-producing cells. In these cells the motion of charge is related to the motion of atoms.

You know that electric charge moves freely in the electric wiring of buildings and in the connecting wires of your laboratory apparatus. Can this flow of charge through solid wires be related to the motion of atoms? It hardly seems possible. In contrast to gases and liquids, solids keep their shape. This means that the atoms of a solid are not free to move about.

To get some idea of what may carry charges in a metal, we shall first describe and analyze an experiment in which charge flows through a vacuum. Figure 12.8(*a*) shows a vacuum tube of a kind once used in radios and TVs. Nearly all the air (about 99.99999%) has been removed from the tube. Figure 12.8(*b*) shows the inside of the tube. The two most important parts are the large, rectangular electrode commonly called the *plate* or *anode*, and the long, narrow cylindrical electrode called the *cathode*. The long zigzag wire is an electric heater that is folded up inside the cathode and is insulated from it (Figure 12.9). When supplied with current, the heater and the surrounding cathode become red hot.

We first examine whether charge can pass through the vacuum tube when the heater is not in use. Figure 12.10 shows the arrangement, including the reading of the ammeter when all connections have been made. The cathode of the tube is connected to the negative end of a 12-cell battery. The plate is connected through an ammeter to the positive end. No current is registered by the ammeter. Next we interchange the connections of cathode and plate. As shown in Figure 12.11, the cathode is connected through the ammeter to the positive end of the battery, and the plate is connected to the negative end. Again there is no current.

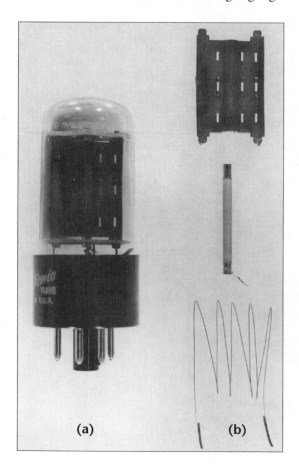

(a) (b)

Figure 12.8

(a) A vacuum tube. (b) The heater (bottom), cathode (center), and plate (top) inside the tube.

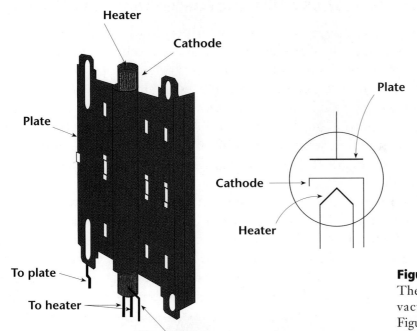

Figure 12.9
The construction of the vacuum tube shown in Figure 12.8, and its schematic representation.

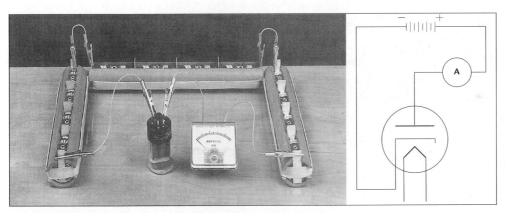

Figure 12.10
Schematic drawing photograph and of the vacuum-tube circuit. The cathode leads to the negative terminal of the power supply.

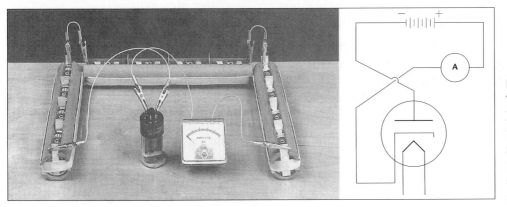

Figure 12.11
The same circuit as in Figure 12.10, except that the cathode is connected to the positive terminal of the power supply (through the ammeter).

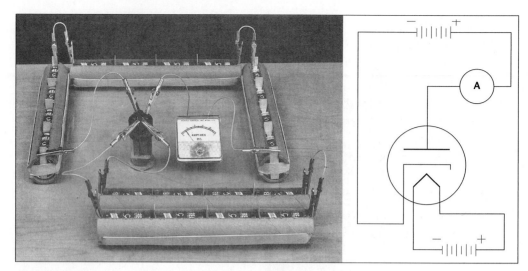

Figure 12.12
The same as Figure 12.10, except that the cathode is heated.

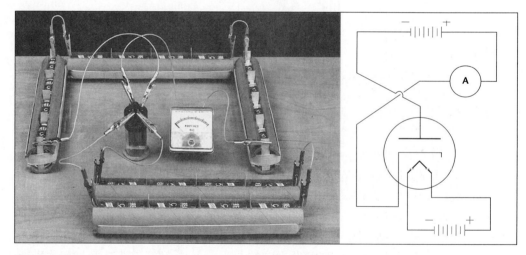

Figure 12.13
The same as Figure 12.11, except that the cathode is heated.

Now we repeat the connections of Figures 12.10 and 12.11, but this time the heater inside the cathode is connected to a separate battery. There is now a current of about 0.15 A when the hot cathode is connected to the negative terminal of the battery (Figure 12.12). But there is no current when it is connected to the positive terminal (Figure 12.13).

We can summarize the behavior of the vacuum tube under the conditions shown in Figures 12.10–12.13 this way: When the cathode is cold, there is no current. When the cathode is heated, there is a current only if the cathode is connected to the negative terminal of the battery.

Does this tell us anything about the direction of motion of the charge in the tube? Does the heating of the cathode enable it to receive charges coming from the plate or to emit charges, which then move to the plate? To help choose between these two possibilities, think of the evaporation of water near the freezing and boiling points. Water evaporates very slowly near the freezing point. It evaporates readily near the boiling point. Apparently, only a few molecules leave the surface of the liquid near the freezing point. Many molecules leave the surface near the boiling point.

This analogy suggests that charges escape from the hot cathode and move to the anode. Does this mean that some of the atoms that make up the cathode actually carry the charge to the plate?

†8. Suppose a vacuum tube has terminals that connect to the cathode, the plate, and the two ends of the heater. If you find which two go to the heater, how can you find out which terminal connects to the plate?

9. If the plate, instead of the cathode, is heated in the cases illustrated in Figures 12.10 and 12.11, what result would you expect to observe in each case?

12.6 Electrons

We can answer the question at the end of the preceding section by massing the cathode of a vacuum tube before it is assembled and again after it has been in use for some time. But we do not even have to do that. Vacuum tubes are mass-produced, so that the mass of the cathode is nearly constant for all tubes of a given type. It is enough, therefore, to compare the mass of the cathode of a new tube with that of a used tube. We did that and found that the mass of the cathode of the new tube and the mass of the cathode of a tube that had run at a current of 0.15 A for 100 hours were the same: 0.224 g for both cases. What decrease in mass should we have expected if the charge flowing through the tube were carried by atoms?

There are 3.6×10^5 seconds in 100 hours, so the total charge that would flow would be $0.15 \text{ A} \times 3.6 \times 10^5 = 5.4 \times 10^4 \text{ A·s}$. This is equal to a flow of

$$\frac{5.4 \times 10^4 \text{ A·s}}{1.60 \times 10^{-19} \text{ A·s/el. ch.}} = 3.4 \times 10^{23} \text{ elementary charges.}$$

Now, to know how much mass is associated with this number of elementary charges, we must know what kind of atoms make up the surface of the cathode. Actually, the cathode surface is composed of a mixture of several oxides; a hot surface composed of this mixture has been found to be a better emitter of charges than a pure metal surface. The lightest of the atoms in these oxides are the oxygen atoms. For simplicity, we shall assume that oxygen atoms carry all the charge that flows.

We shall further assume that one oxygen atom carries two elementary charges, the same charge that is required to liberate an oxygen atom in the electrolysis of water. Thus

$$\frac{3.4 \times 10^{23} \text{ el. ch.}}{2 \text{ el. ch./atom}} = 1.7 \times 10^{23} \text{ atoms}$$

of oxygen would leave the cathode in 100 hours. Since one atom of oxygen has a mass of 2.7×10^{-23} g, the loss in mass of the cathode should be

$$1.7 \times 10^{23} \times 2.7 \times 10^{-23} = 4.6 \text{ g.}$$

This is more than twenty times the entire mass of the cathode!

The conclusion is inescapable: the charge is not carried across the vacuum tube by atoms. There must be carriers of a different kind which, from our observations, have the following properties:

(a) They come off a hot cathode connected to the negative end of a battery but *not* from a hot cathode connected to the positive end.

(b) They do not seem to reduce the mass of the cathode.

These carriers are called *electrons*. Since they move from a cathode that is connected to the negative end of a battery, they are said to be *negatively charged*.

---

10. Suppose the cathode of a vacuum tube is made of copper. What would be the loss in mass after 100 hours at 0.15 A if copper atoms carried the charge across the tube?

11. The mass of an electron is about 1/2,000 of the mass of a hydrogen atom. How long would you have to run the vacuum tube described in Section 12.5 in order to show that electrons do not accumulate on the plate, increasing its mass at the expense of the cathode?

12.7 Atoms and Ions

Very early in our study of electricity we adopted the eighteenth-century idea that when an electrical device is operating, something is moving through it. We called that something the electric charge and developed ways of measuring it (Section 10.1). Now we are in a position to develop this idea into a more concrete model. In so doing we shall also bring the model up to date.

The first thing the model must accomplish is to tie together the motion of charge when it is accompanied by the motion of atoms and its motion when it is not. Consider a circuit consisting of a battery, a copper-plating cell, and a vacuum tube (Figure 12.14). You know from experience that in the copper cell, copper dissolves at the electrode leading to the positive terminal of a battery and plates out on the elctrode leading to the negative terminal. Charge carriers that move away from the electrode connected to the positive terminal of the battery are called *positively* charged. In the vacuum tube the charge carriers are electrons, which move away from the heated cathode when it leads to the negative terminal of the battery. Electrons are *negatively* charged.

The problem now facing us is: how can electrons and atoms be related to account for the overall motion of charge in the circuit? This question was answered satisfactorily only in the twentieth century. Although the resulting model was based on a variety of experiments, the passage of charge through cells, metal wires, and vacuum tubes provides the main clues.

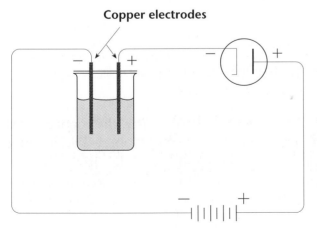

Copper electrodes

Figure 12.14
A circuit consisting of a battery, a vacuum tube, and a copper-plating cell. The heating circuit for the cathode is not shown.

The basic features of the model are as follows: An atom is made up of a part that is positively charged and a number of negatively charged electrons. The total charge on an isolated atom is normally zero. Since the atom has no net charge, it is electrically neutral. If an atom loses an electron, it becomes positively charged. On the other hand, if an atom gains an electron, it becomes negatively charged. Charged atoms are called *ions*.

For example, when a neutral sodium atom loses an electron, it becomes a positive sodium ion. This is expressed in symbols as

$$Na \rightarrow Na^+ + e^-.$$

Similarly, when a neutral chlorine atom gains an electron, it becomes a negative chlorine ion:

$$Cl + e^- \rightarrow Cl^-.$$

Copper and zinc atoms carry two elementary charges in a solution. Thus each of them must give up two electrons to become a doubly charged ion.

$$Cu \rightarrow Cu^{++} + 2e^-$$

$$Zn \rightarrow Zn^{++} + 2e^-$$

We shall assume that in a solution that conducts electricity, the solute is in ionic form. For example, a solution of copper chloride, $CuCl_2$, conducts electricity. We can consider the formation of the ions as the combination of the two steps

$$Cu \rightarrow Cu^{++} + 2e^-$$

and

$$2Cl + 2e^- \rightarrow 2Cl^-$$

occurring together.* The net result is

$$CuCl_2 \rightarrow Cu^{++} + 2Cl^-.$$

When copper sulfate, $CuSO_4$, is dissolved in water, the copper atoms separate from the sulfur and oxygen atoms as Cu^{++} ions. The negative ions, called *sulfate ions*, are each made up of one sulfur atom and four oxygen atoms tightly bound together. We do not know how the charges are distributed among the sulfur atom and the four oxygen atoms.

---

*"$2Cl^-$" means two separate chlorine ions, whereas the subscript 2 in $CuCl_2$ means that in this chloride of copper there are two chlorine atoms for every copper atom.

Therefore, we shall simply write SO_4^{--} for the sulfate ion to indicate that two electrons are attached to the sulfate. When an electric current passes through the copper-plating cell in Figure 12.14, copper from the positive electrode enters the solution as Cu^{++} ions, and copper from the solution plates out on the negative electrode as neutral copper atoms. In terms of ions and electrons this suggests the following reactions:

At the positive electrode, $Cu \rightarrow Cu^{++} + 2e^-$

At the negative electrode, $Cu^{++} + 2e^- \rightarrow Cu.$

12. Suppose a copper-plating cell and a vacuum tube were connected in series with a battery. When the heater of the tube was connected, charge moved through the circuit. Do the copper ions in the cell move away from the same battery terminal as the electrons in the tube?

†13. A Daniell cell is used to run a copper-plating cell, as shown in Figure A.

Figure A
For problem 13

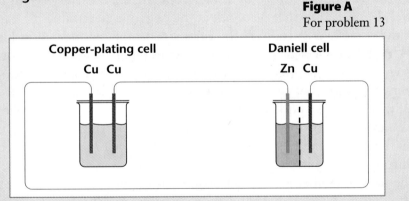

a. In which direction do the copper ions move in the copper cell? In the Daniell cell?

b. In which direction do the zinc ions move in the Daniell cell?

c. In which direction do electrons move in the wires?

14. When a piece of zinc is dipped into a solution of copper sulfate, zinc dissolves and copper precipitates. Describe this process in terms of atoms, electrons, and ions.

15. What is the number of elementary charges carried by an electron? Base your answer on your plating experiments and the ion-electron model of the atom.

12.8 The Motion of Charge Through a Circuit

In the copper-plating cell in Figure 12.14, positive copper ions carry charge through the solution. How is charge carried through the metal wires?

From the reaction

$$Cu \rightarrow Cu^{++} + 2e^-$$

at the positive electrode, it follows that electrons are moving away from this electrode. By the same reasoning, we conclude from the reaction

$$Cu^{++} + 2e^- \rightarrow Cu$$

at the negative electrode that electrons move from the wire to the electrode. To complete the model we must assume that the electrons can move through metals while ions cannot.

Now we can use the basic features of the model to describe what happens on the atomic level when charge moves all the way around the circuit shown in Figure 12.14. Let us begin at the negative terminal of the battery. Electrons move in the wire from this terminal to the copper electrode on the left side of the copper-plating cell. There they combine with positive copper ions that come from the solution. This produces neutral copper atoms that plate out at the negative electrode.

At the positive electrode on the right side of the plating cell in Figure 12.14, neutral copper atoms dissolve, giving up electrons and becoming positive copper ions. At this electrode, copper dissolves and replaces the copper that plates out of the solution on the left. Copper ions are continually being removed from the solution at the left-hand electrode and at the same time being supplied to the solution at the right-hand electrode. Therefore there is a motion of positive copper ions from right to left through the solution.

The neutral copper atoms that dissolve at the positive electrode of the plating cell give up electrons. These electrons move from the positive electrode through the wire to the hot cathode of the vacuum tube. There they "boil off" into the vacuum and move across to the plate and through the wire to the positive terminal of the battery.

This model, describing the flow of charge on the atomic level, is consistent with charge conservation in the circuit. Electrons move from the battery to a copper-plating cell. An equal number of electrons move from the copper-plating cell through the vacuum tube and then back to the battery.

What happens in the battery itself? The details depend on the kind of cells that make up the battery. But all battery cells have one property in common: they have the ability to push electrons away from one electrode and accept them at the other.

12.9 The Direction of Electric Current

The motion of charge in the copper-plating cell is particularly simple. Copper dissolves at one electrode and plates out at the other, while the rest of the cell remains unchanged. Therefore charge can be carried across only by positive copper ions. If we replace the copper cell in Figure 12.14 by another cell consisting of two carbon electrodes and a solution of hydrochloric acid, the situation will be more complex.

Although the electrodes themselves will remain unchanged, hydrogen will evolve at the electrode leading to the negative terminal of the battery. Also, chlorine will evolve at the electrode leading to the positive terminal. This indicates that positive hydrogen ions move to the negative electrode, and negative chlorine ions move to the positive electrode. What, then, is the direction of the current in the circuit? Is it the direction of the positive ions in the solution or the direction of the negative ions in the solution and the electrons in the wire?

It appears that we could choose either direction for the direction of the current. In fact, the direction of the current is commonly taken to be the same as the direction of motion of the positive ions. The direction of the current and the direction of motion of the ions and electrons is shown in Figure 12.15.

In Figure 12.14 the direction of the current is counterclockwise. This is the direction of the positive ions in the copper cell. It is opposite to

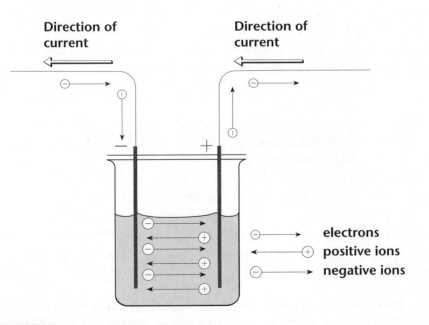

Figure 12.15
The direction of electric current through the cell in Figure 12.14 as related to the motion of ions and electrons.

the direction of motion of the electrons in the vacuum tube and in the wires. Notice that outside the battery the direction of the current is from the positive to the negative terminal. However, inside the battery it is from the negative to the positive terminal.

FOR REVIEW, APPLICATIONS, AND EXTENSIONS

16. An experiment calls for running a Daniell cell for 15 minutes each day for a week. How would you turn off the cell at the end of each day's run so that you could continue to use the same materials for the cell?

17. A Daniell cell can be made without a divider, as shown in Figure B.
    a. Why do the solutions not mix immediately?
    b. If the cell is allowed to stand unused, what will eventually happen?
    c. Such cells were used commercially in the early days of telegraphy. However, they were never allowed to stand without producing at least a little current. Why was this?

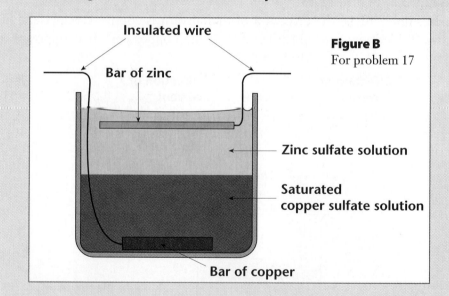

Insulated wire

Bar of zinc

Figure B
For problem 17

Zinc sulfate solution

Saturated copper sulfate solution

Bar of copper

18. Will a Daniell cell produce current if both electrodes are made of zinc? Of copper?

19. Recall what happened when you dipped two zinc electrodes connected by an ammeter in a solution of copper sulfate. What do you think the ammeter would read if copper plated out uniformly and at the same rate on both electrodes without falling off?

20. If you were to make an iron-copper cell, which electrode would you expect to be negative?

21. A circuit like that in Figure 12.12 is wired up with an additional ammeter placed in the heater circuit. Before the battery at the top is connected, the heater is turned on and a current of 0.60 A is read on the meter in the heater circuit. When the battery at the top is connected, the meter at the top in the cathode-plate circuit reads 0.16 A. What reading would you now expect to find on the heater-circuit meter?

22. You have worked with hydrogen cells, plating cells, ammeters, light bulbs, and vacuum tubes. If you place these things in a circuit, which ones will enable you to tell the "+" terminal of an unmarked battery? How would you tell?

23. In some types of vacuum tube, the heater is used as the cathode. Such a tube is shown in Figure C. On this ammeter, the zero position of the needle is in the center of the scale, and the needle can move in either direction. S_1 and S_2 are switches.

    What do you predict you would observe if the circuit were operated under each of the following conditions?

	Switch S_1	Switch S_2
a.	Open	Closed upward
b.	Closed	Closed upward
c.	Closed	Closed downward

Figure C
For problem 23

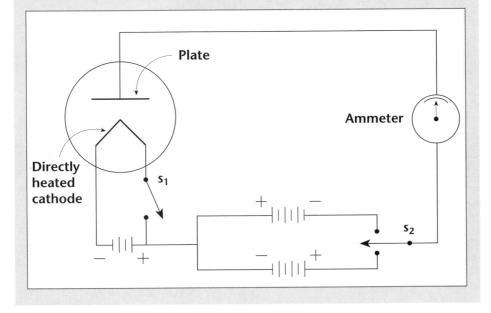

24. Suppose you have a cell consisting of two copper plates and a solution of sulfuric acid. What do you predict will happen if you pass a current through this cell for some time?

25. A student using the evidence so far in the course proposed that ions form only when water is present. Does the evidence presented in Section 6.10 support this argument?

THEME FOR A SHORT ESSAY

Vacuum tubes were invented early in this century. They played a crucial role in the development of radios and later, computers. Today they are hardly used anymore, having been replaced by transistors and integrated circuits. Other inventions, such as the steam locomotive and dirigible, have also played an important role in the development of technology only to be replaced by newer inventions. Read about an invention of your choice and write a paper titled "The Rise and Fall of . . . ."

Epilogue

As this course comes to an end, you may ask yourself: "What have I learned this year in science?" We hope that you will think of several things, some specific and some general.

During the year we tried to familiarize you with some of the basic facts and ideas of physical science. You saw the evidence for the facts, and the usefulness of the ideas. Contrary to what you may have expected, science does not deal with absolute truths. The specific facts we find in the laboratory, such as melting points, solubilities, half-lives, and charge per atom, are all subject to the limitations of our measurements.

Scientific laws are generalizations based on measurements made many times under controlled conditions to assure their validity. Yet these generalizations also have their limitations. If this is the case in science, how careful must you be about the facts and generalizations in your daily life? Do you ask for evidence to support what you read and hear? If this introduction to science has made you a more critical reader, a more careful observer, and a sharper thinker, your work during the year was worthwhile.

Answers to Questions Marked with a Dagger (†)

Chapter 1

3. a. 8 cubes
   b. 27 cubes
   c. 8 cm^3, 27 cm^3
7. a. 0.1 cm^3
   b. 0.2 cm^3
11. a. 10 cm^3
    b. 20 cm^3
    c. 20 cm^3
    d. 30 cm^3
    e. 0.60
17. 18.325

Chapter 2

1. Boil away or evaporate the water. The mass of the salt recovered would be the same as the beginning.
2. No
9. a. 2.956 g
   b. 0.054 g
   c. About 2 percent

Chapter 3

2. a. A
   b. C
   c. B
14. 1.7 g/cm^3
16. a. 10.5 g/cm^3
    b. 2.1 g/cm^3
    c. 0.82 g/cm^3
18. a. 0.50 g
    b. $1.1 \times 10^{-3} \text{g/cm}^3$

27. a. Gas
    b. Solid
    c. Solid or liquid
    d. Solid or liquid
    e. Gas

Chapter 4

10. From Figure 4.3: 47°C
12. From Figure 4.3: 212 g
18. The solubility of oxygen in water decreases as the temperature of the water rises.

Chapter 5

1. They must have different boiling points.
6. They must differ greatly in solubility.

Chapter 6

2. a. 180 g
   b. 180 g
3. No
5. a. 18cm^3
   b. $2.4 \times 10^4 \text{ cm}^3$
9. Tube I: 25 cm^3 hydrogen
   Tube II: 25 cm^3 oxygen
   Tube III: 0 cm^3
13. Ratios: b, c, d, f, g

Chapter 7

1. Y
4. 1.0×10^2 counts per minute

11. There is no easy way to tell, although all three directions are possible. One could assume that the shorter tracks either hit the bottom or came up out of the cloud forming region.

Chapter 8

1. It must suggest at least one new experiment and correctly predict the results.
5. No

Chapter 9

1. $10^7 \, cm^2$
3. $1.4 \times 10^{-2} \, g$
21. $100 \, cm^3$
23. 1.3×10^{18}. We assumed that the disintegration of one polonium atom produces one helium atom and one lead atom.

Chapter 10

1. a. Indirectly
   b. Directly
   c. Indirectly
4. a. Yes
   b. 1/1
9. $10 \, cm^3$
12. The charges are equal.
15. a. 20 ampere-seconds
    b. 72 ampere-seconds
    c. 72 ampere-seconds

Chapter 11

1. The test tube in which hydrogen is collected
11. One elementary charge

13. So little charge would pass through the solution that no change in mass could be measured.

Chapter 12

8. Connect an ammeter and battery in series between the two remaining connections. Connect the heater circuit. If the ammeter shows a current, then the positive end of the battery is connected to the plate terminal.
13. a. Left to right in both cells
    b. Left to right
    c. Counterclockwise

List of Tables

Acknowledgments from the First Edition

The following members of the Educational Services Incorporated staff in addition to myself were involved in the development of the course; John B. Coulter, on leave from Pakuranga College, Howick, New Zealand; Judson B. Cross; John H. Dodge; Robert W. Estin, on leave from Roosevelt University, Chicago, Illinois; Malcolm H. Forbes; Ervin H. Hoffart; Gerardo Melcher, on leave from the University of Chile, Santiago, Chile; Harold A. Pratt, on leave from Jefferson County Public Schools, Lakewood, Colorado; Louis E. Smith, on leave from San Diego State College, San Diego, California; Darrel W. Tomer, on leave from Hanford Union High School, Hanford, California; and James A. Walter.

For the summer of 1963, we were joined by Elmer L. Galley, Mott Program of the Flint Public Schools, Flint, Michigan; Edward A. Shore, The Putney School, Putney, Vermont; and Byron L. Youtz, Reed College, Portland, Oregon.

Later on, considerable time was devoted to this project by others who joined us during the summers or consulted on a part-time basis throughout the following years; Gilbert H. Daenzer, Lutheran High School Central, St. Louis, Missouri; Thomas J. Dillon, Concord- Carlisle High School, Concord, Massachusetts; Winslow Durgin, Xavier High School, Concord, Massachusetts; Alan Holden, Bell Telephone Laboratories, Murray Hill, New Jersey; Robert Gardner, Salisbury School, Salisbury, Connecticut; Father John Kerdiejus, S.J. Xavier High School, Concord, Massachusetts; Herman H. Kirkpatrick, Roosevelt High School, Des Moines, Iowa; Elisabeth Lincoln, Dana Hall School, Wellesley, Massachusetts; John V. Manuelian, Warren Junior High School, Newton, Massachusetts; John N. Meade, Newman Junior High School, Needham, Massachusetts; Paul Meunier, Marshfield High School, Marshfield, Massachusetts; Father Patrick Nowlan, O.S.A., Monsignor Bonner High School, Drexel Hill, Pennsylvania; Frank Oppenheimer, University of Colorado, Boulder, Colorado; Charles M. Shull, Jr., Colorado School of Mines, Golden Colorado; Malcolm K. Smith, Massachusetts Institute of Technology, Cambridge, Massachusetts; Moddie D. Taylor, Howard University, Washington, D.C.; Carol A. Wallbank, Dighton-Rehoboth Regional High School, Rehoboth, Massachusetts; Richard Whitney, Roxbury Latin School, West Roxbury, Massachusetts; Marvin Williams,

Bell Junior High School, Golden, Colorado; M. Kent Wilson, Tufts University, Medford, Massachusetts; and Carl Worster, Belmont Junior High School, Lakewood, Colorado.

I also wish to acknowledge the invaluable services of George D. Cope and Joan E. Hamblin in photography; R. Paul Larkin as art director for the preliminary edition; Barbara Griffin, Nancy Nelson, and Gertrude Rogers in the organization of feedback from the pilot schools; Nathaniel C. Burwash and John W. DeRoy in apparatus construction and design; Benjamin T. Richards for production; and Andrea G. Julian for editorial assistance. Much of the administrative work was done by Geraldine Kline.

Throughout the entire project I benefited from the advice and criticism of M. Kent Wilson. Valuable assistance in coordinating various group efforts in the summers of 1963 and 1966 was provided by Byron L. Youtz. In editing this edition of the course, I was specially aided by Judson B. Cross and by Harold A. Pratt, who was responsible for the group summarizing the feedback.

I wish to thank the editorial and art staff of the Educational Book Division of Prentice-Hall, Inc., for their help in preparing the final form of this edition.

Constant sources of encouragement and constructive criticism were the pilot teachers, who voluntarily spent many extra hours relating to us their classroom experience. Without them, the course could not have been developed to this point.

The initial stage of the Introductory Physical Science Program was funded by Educational Services Incorporated. Since then, it has been supported by a grant from the National Science Foundation. This financial support is gratefully acknowledged.

Uri Haber-Schaim
March 1967

Index

F

mass, 14–5, 40

    change in, 28

    conservation of, 36–7

    standard, 15

mass ratio. *See* law of constant proportions

melting point

    as characteristic property, 59–60

measuring, 42–3

mercury

    chlorides of, 223

    density of, 173

    in drinking water, 85

    oxides of, 107, 125, 156, 223

methane (marsh gas), 78

methanol, 72–4

mg (milligram), 16

microburner, use of, 3–4

mixture, definition, 107

mL (milliliter), 8

mm (millimeter), 9

models, 152, 154–155, 169

molecules, 160

motion of charge. *See* current, electric

multiple proportions, law of, 222–3

muriatic acid, 126 (*see also* hydrochloric acid)

N

negative charge, 241

nickel, electroplating, 216

nitrogen, 103

    oxides of, 160

    solubility of, 82

notation, scientific. *See* scientific notation

nuclear waste, 170

O

objects, distinguished from substances, 40

oil of vitriol (sulfuric acid), 76 (*see also* sulfuric acid)

oleic acid, 179–82

ore, 129–31

oxygen, 103, 128–9, 160

    solubility of, 82

P

paper chromatography, 101–102, 110

paraffin, 96

Pasteur, Louis, 169

pentane, 96

petroleum, 93–6, 107

plate. *See* anode

polonium, 139, 168, 170

    atomic mass of, 186–8

    counting rate, graph, 145

    decay: 182–7; and helium,
        production of, 182–7; in
        solution, 192

    spectrum of, 164

positive charge, 241

potassium

    discovery of, 126

    in electrolytic cells, 218

potassium nitrate

    precipitation of, 109

    solubility of, 71–2, 98–100

potassium sulfate, solubility,
    70

powdered carbon, in batteries,
    231

powers-of-10 notation. *See*
    scientific notation

precipitate, 71

pressure

    and density, 57–8, 63

    and volume, 13

propane

    separating from air, 102

Proust, Joseph Louis, 121

pure substances, 107, 126

Q

quantity of charge, measuring,
    206–8

R

radiation, as food preservative,
    147

radioactive background, 140–1,
    144, 147

radioactive decay 136–50,
    166–8

    and health, 146–7, 150

    rate of, 142–4, 167

radioactive elements, 139–40,
    168

radium, 168

radon, 146, 160, 168

    effect on lungs, 146–150

rock salt, density, 53

rider. *See* balance

rubber stopper, safety with, 3

rubbing alcohol. *See*
    isopropanol

ruler, use of, 9–10

S

salt. *See* rock salt; sodium chloride

sand, volume of, measuring, 12, 25

saturated solution, 66, 68–9

scale, boiler, 86

scales, reading, 9–10

scientific notation, 37, 58, 174–8

　multiplying and dividing with, 176–8

sensitivity of balance, 21–2

separation

　by condensation, 103–4

　by flotation, 97

　of gases in mixtures, 102–3

　of insoluble solids, 97

　of liquids in mixtures, 90–2, 93, 95–7

　of soluble solids in mixtures, 98–100

　see also particular substances

significant digits, 49–50, 177–8

silicon dioxide, 129

silver, density of, 53

sinkholes, 83

sodium

　discovery of, 127

ions of, 242

　spectrum of, 162

sodium carbonate, 113

　separating, 109

sodium chloride (table salt)

　boiling point of, 60

　decomposition of, 217

　dissolving, 133

　precipitating, 109

　purifying, 98

　solubility of, 71–2, 98–100

sodium nitrate, solubility, 71–2

solids

　density of, measuring, 47–9

　internal structure, 159

solubility (*see also* concentration; fractional crystallization; precipitate)

　as characteristic property, 69, 75

　graph, 71

　definition of, 69

　of gases, table, 82

　measuring, 79–81

　and separation of substances, 98–100

　and temperature, 69–72

solute, definition of, 67

Volta, Alessandro, 126
volume
    calculating, 7, 172–3, 179–81
    definition of, 6
    and displacement, 11–12,
        23, 24–5
    measuring: of gases, 13; of
        liquids, 8, 25; of solids,
        11–12, 24–5
    not a characteristic
        property, 40
    and quantity of matter, 7,
        13–14, 37

electroplating, 215–7
    used to prevent corrosion,
        234–5
zinc chloride, synthesis, 117–9

W

washing soda, 69, 87
water
    boiling point of, 49
    decomposition of, 112–3,
        116–7, 133, 214–5 (*see also*
        electrolysis)
    synthesis, 114–6
wood alcohol (methanol), 72–4

Z

zinc
    in battery, 231–2
    in Daniell cell, 228–30
    in galvanized iron, 235